Email us at
schoolhome.kids@gmail.com

to get free extras

just titre the email –MATH KIDS-
And we will send some extra
surprises your way

Trace Number - 1

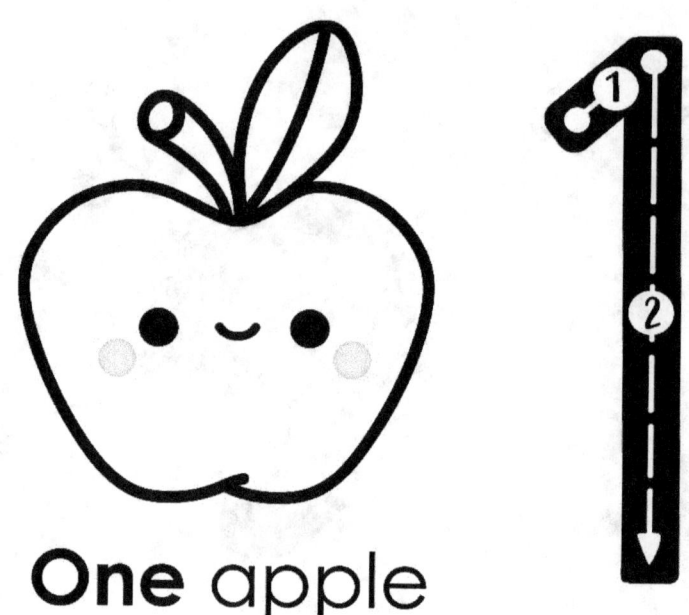

One apple

1 1 1 1 1 1

1 1 1 1 1 1

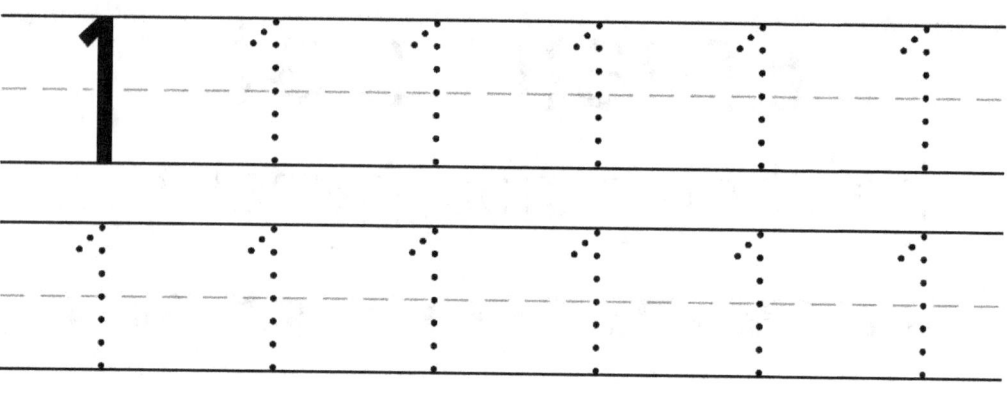

one one one

one one one

Trace Number - 2

Two pineapples

2 2 2 2 2

2 2 2 2

two two two

two two two

Trace Number - 3

Three strawberries

3 3 3 3 3

3 3 3 3 3

three three

three three

Trace Number - 4

Four lemons

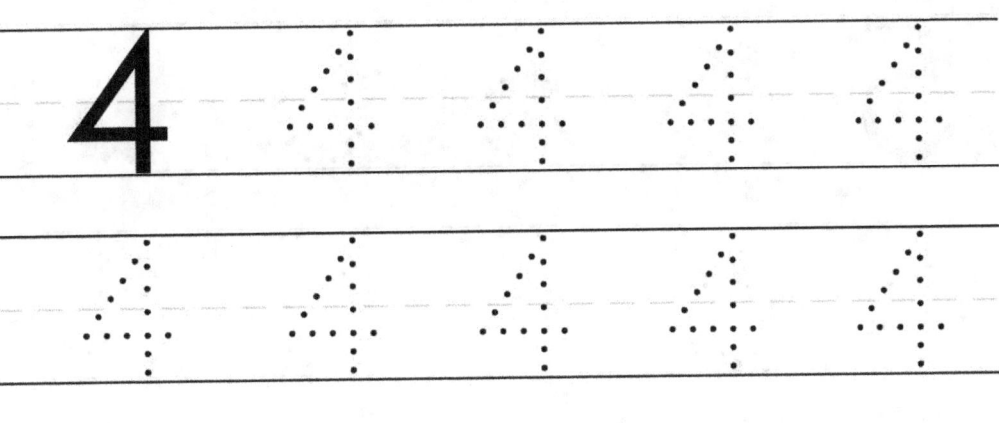

Trace Number - 5

Five oranges

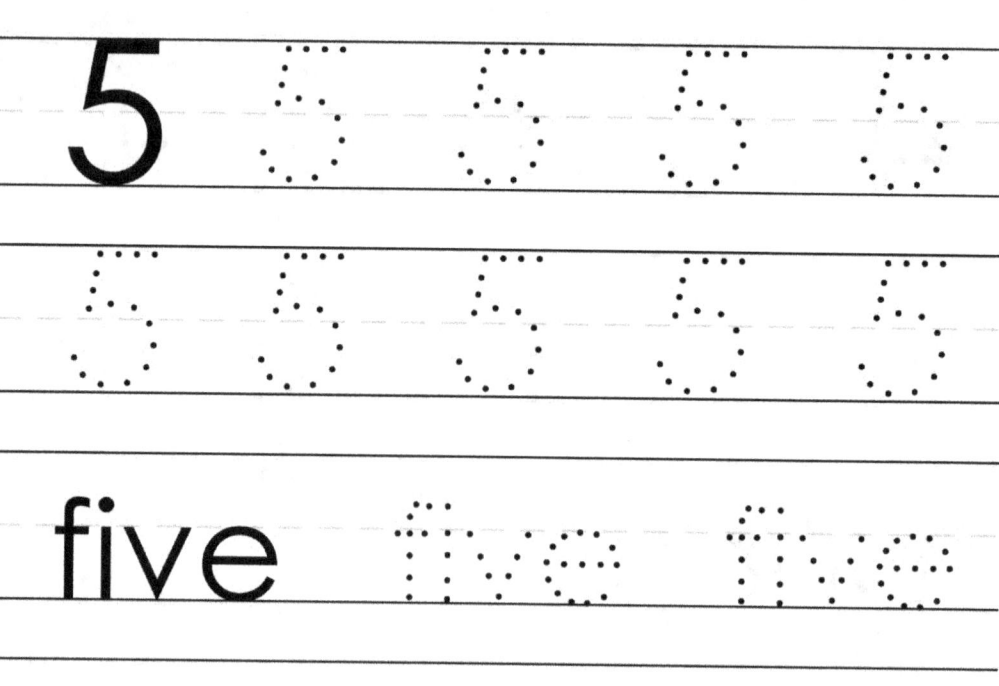

Trace Number - 6

Six broccolis

6 6 6 6 6

6 6 6 6 6

six six six six

six six six six

Trace Number - 7

Seven tomatos

7

seven

Trace Number - 8

Eight eggplants

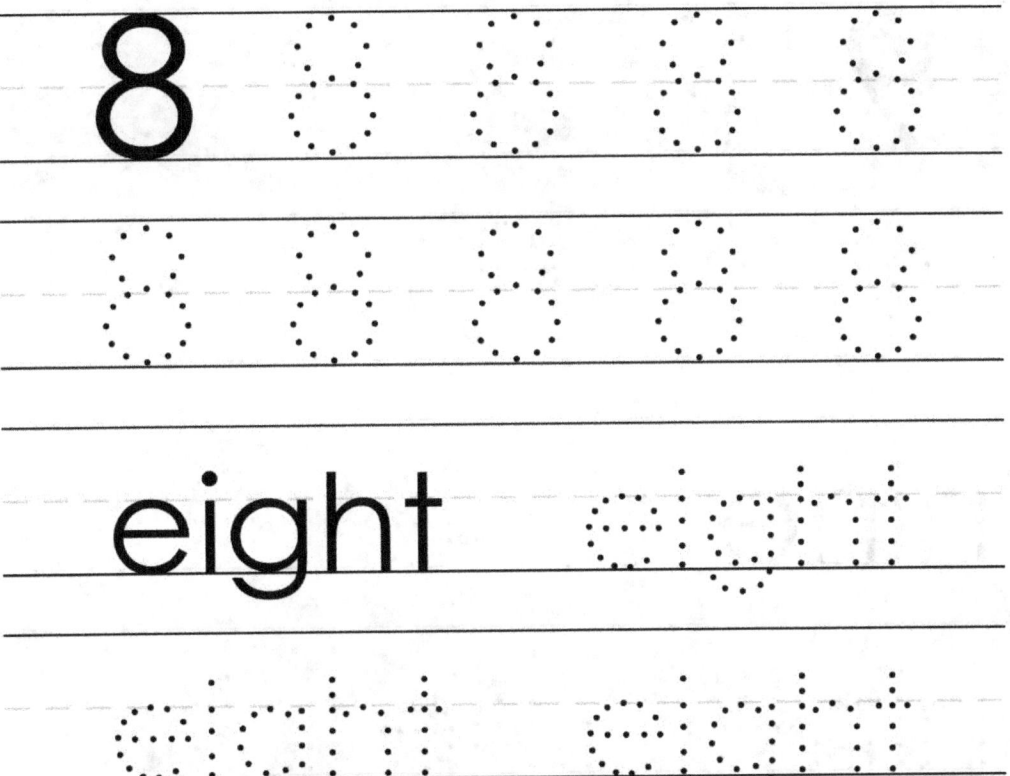

Trace Number - 9

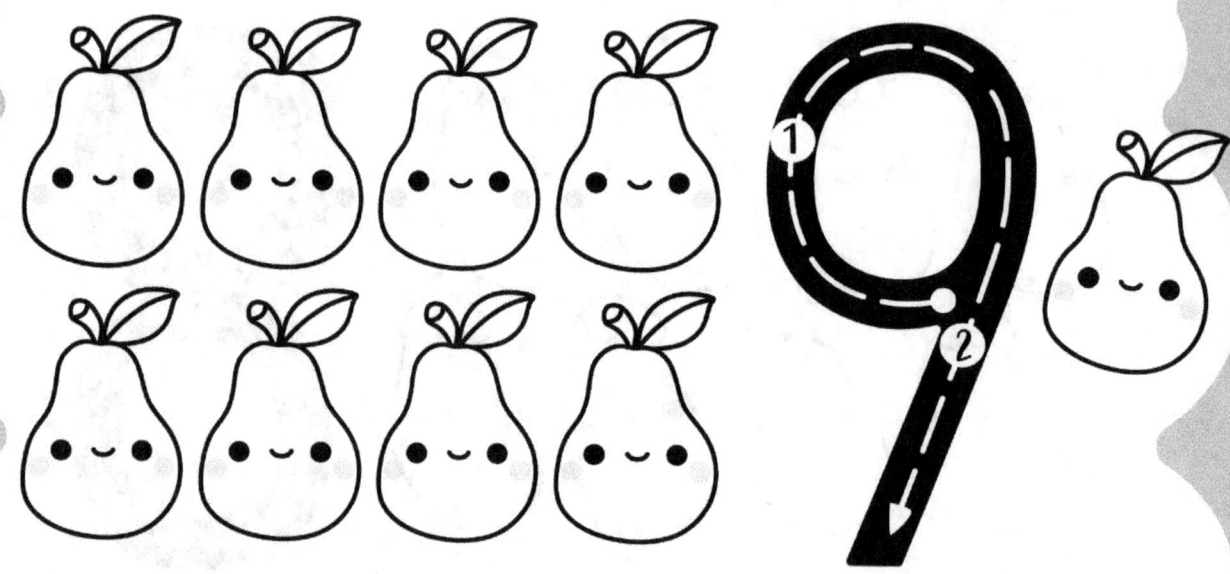

Nine pears

9 9 9 9 9

9 9 9 9

nine nine nine

nine nine nine

Trace Number - 10

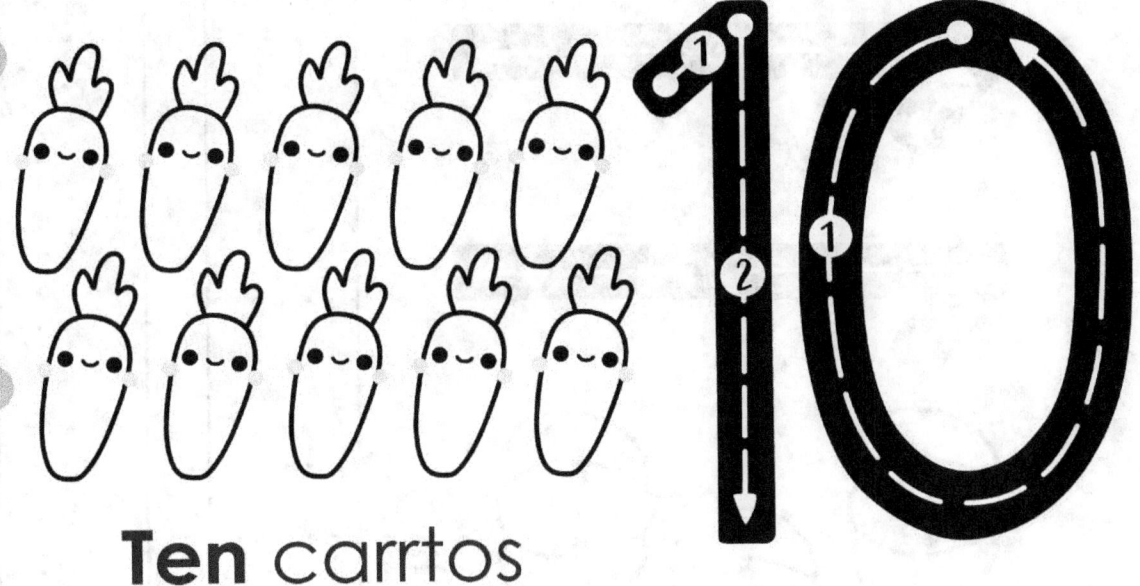

Ten carrtos

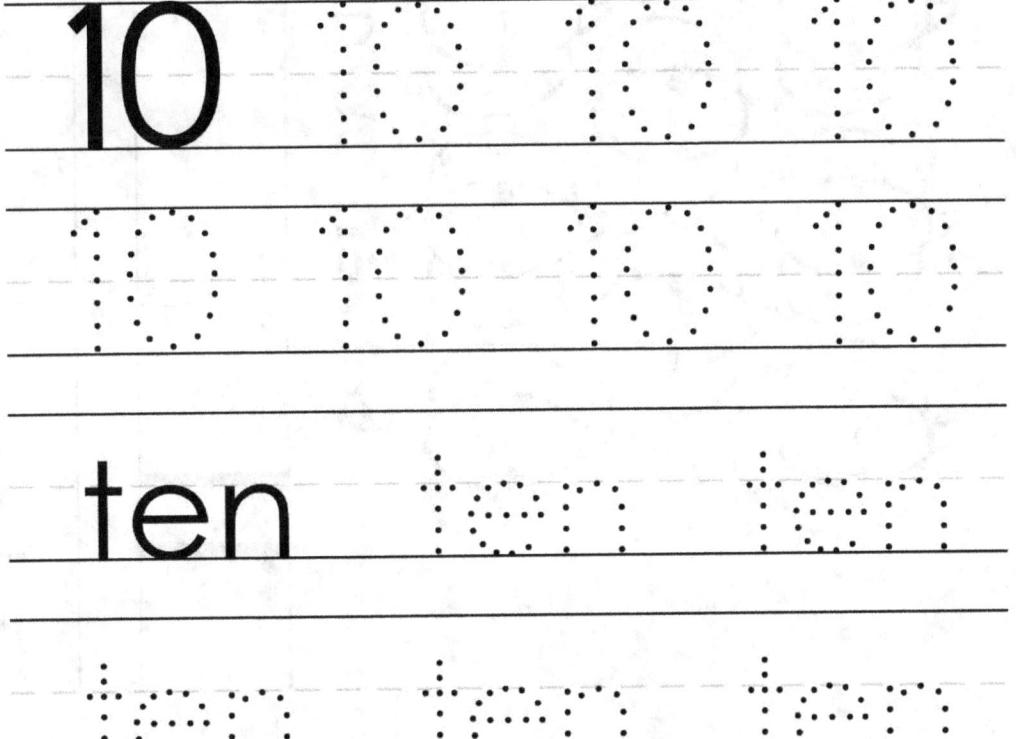

Trace Number - 11

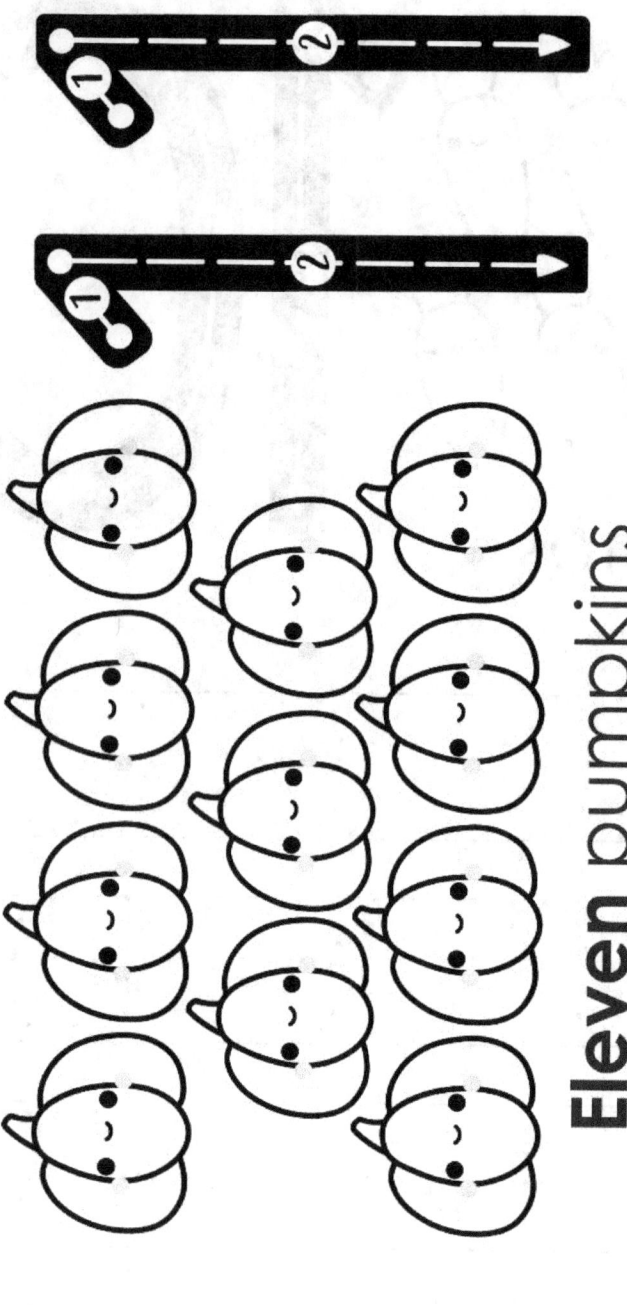

Eleven pumpkins

11

eleven

Trace Number - 12

Twelve beetroots

12

twelve

Trace Number - 13

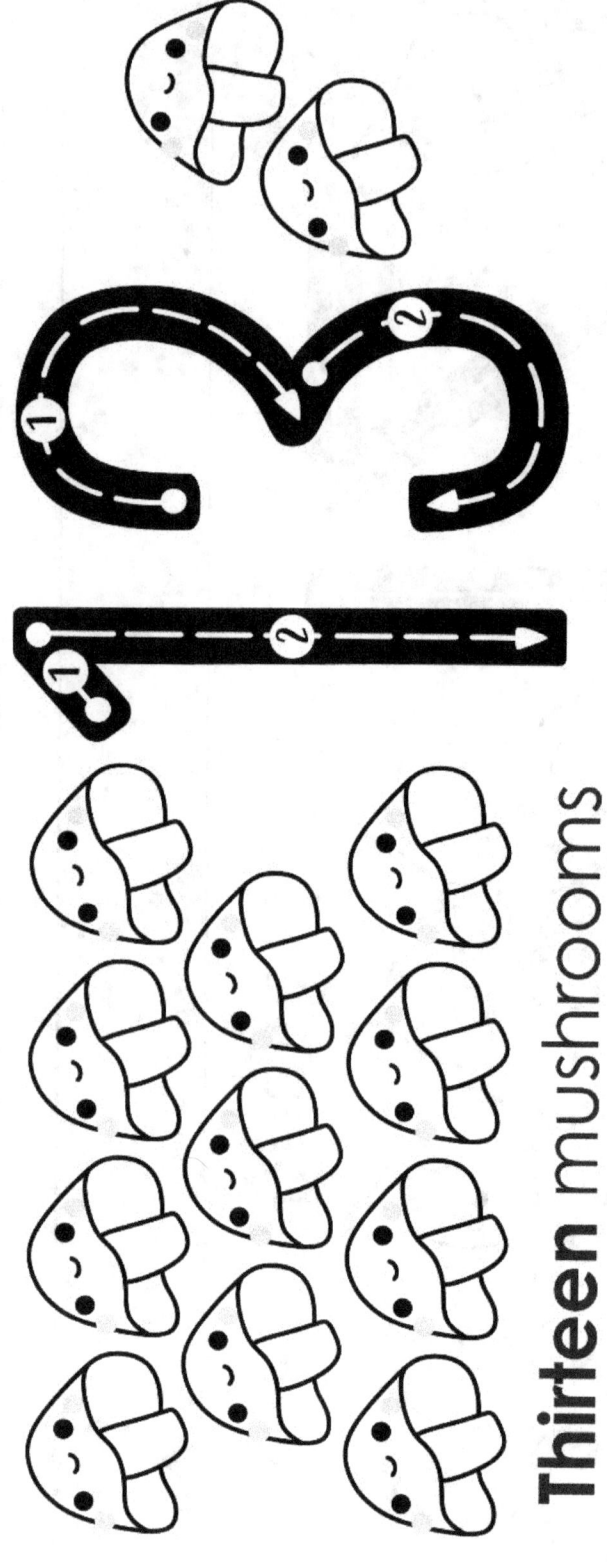

Thirteen mushrooms

thirteen

Trace Number - 14

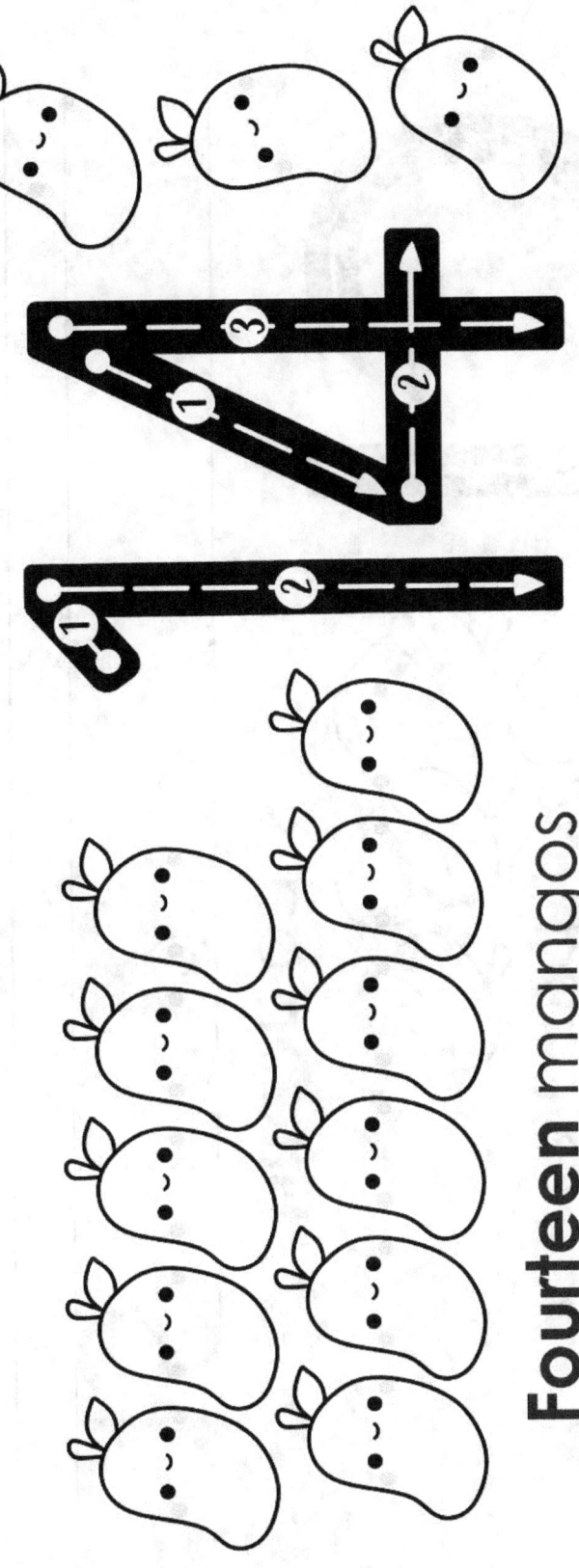

Fourteen mangos

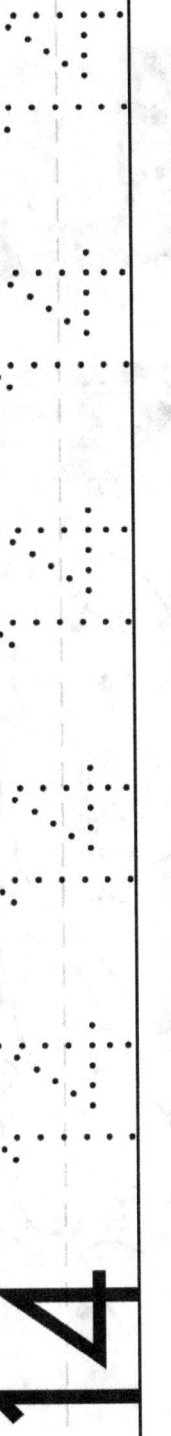

14

fourteen

Trace Number - 15

Fifteen oaks

15

fifteen

Letter Tracing

Letter Tracing
Alphabet Writing Practice

Name:

A is for
Ant

Trace the letters with a pencil. Then practice writing the letters on the lines

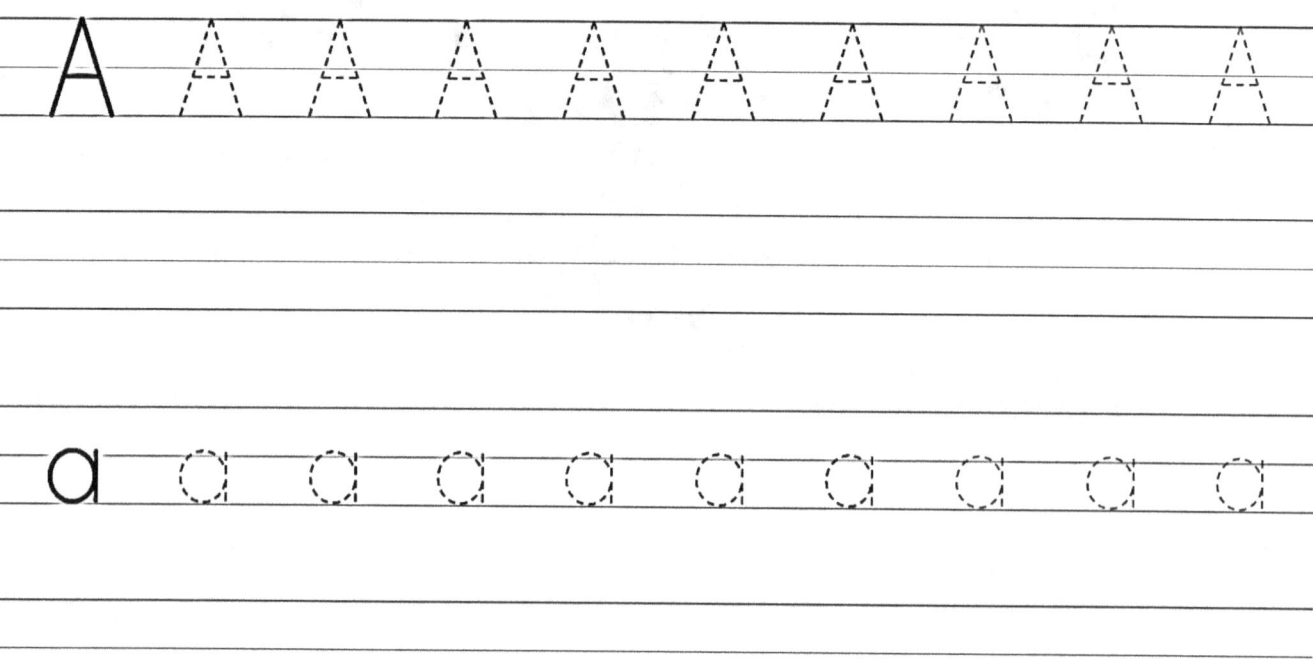

Letter Tracing
Alphabet Writing Practice

Name:

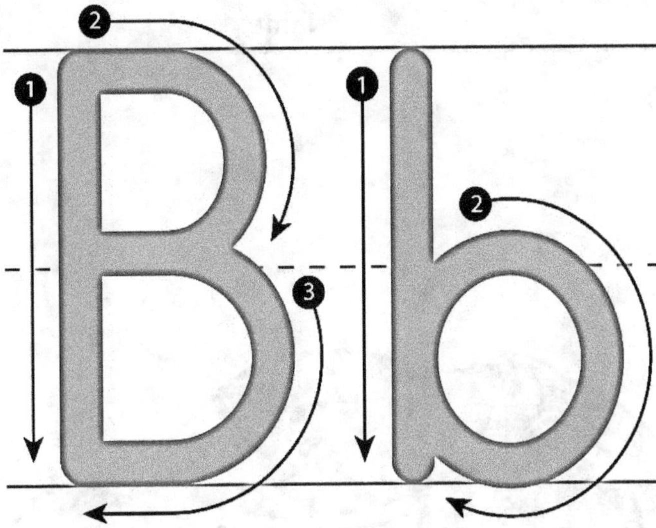

Bb

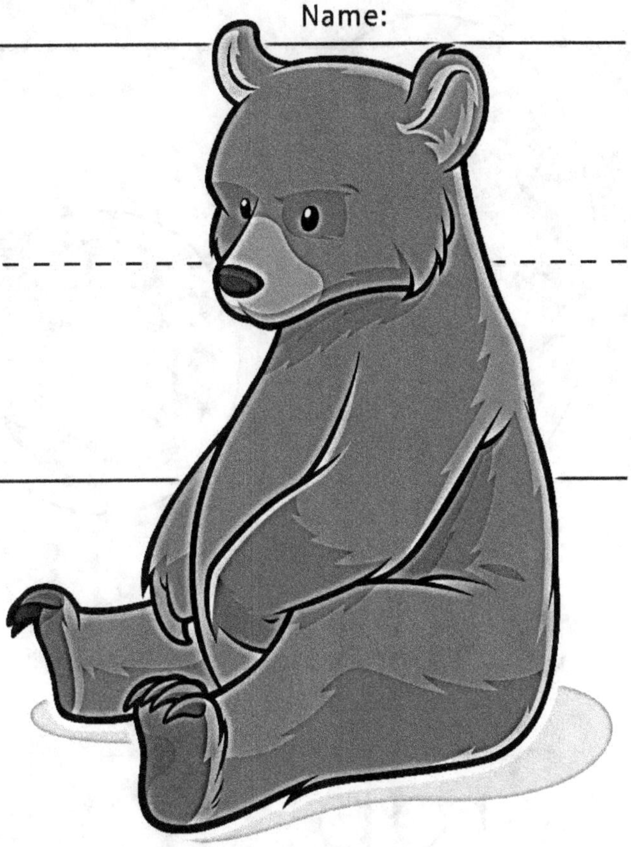

B is for Bear

Trace the letters with a pencil. Then practice writing the letters on the lines

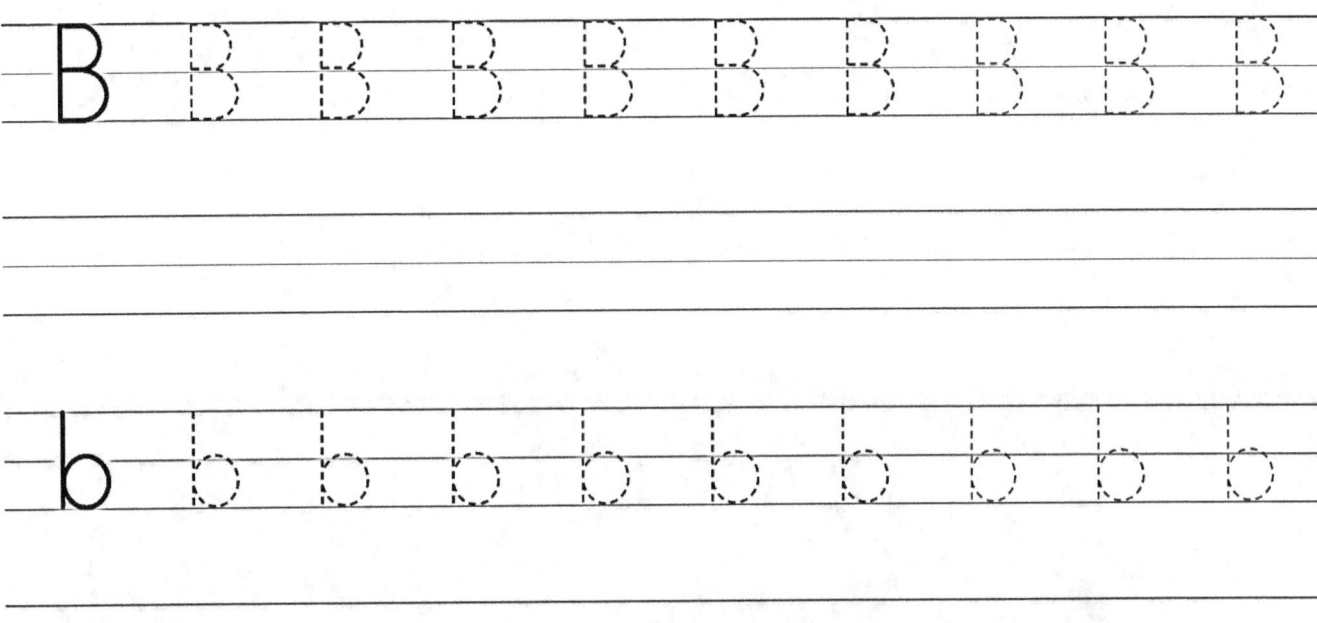

Letter Tracing
Alphabet Writing Practice

Name: _____

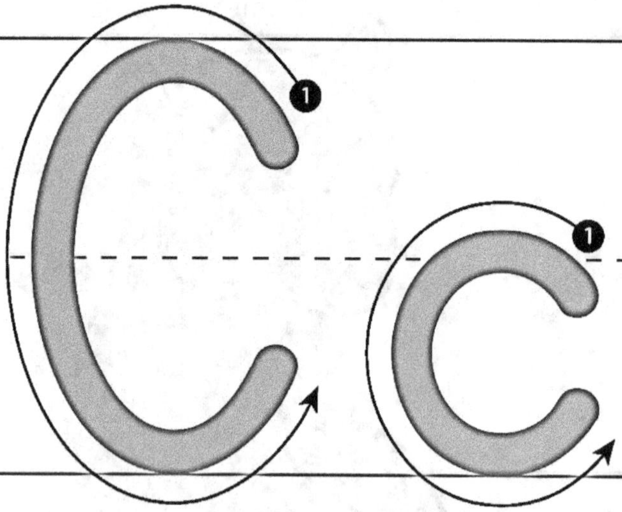

C c

C is for Crab

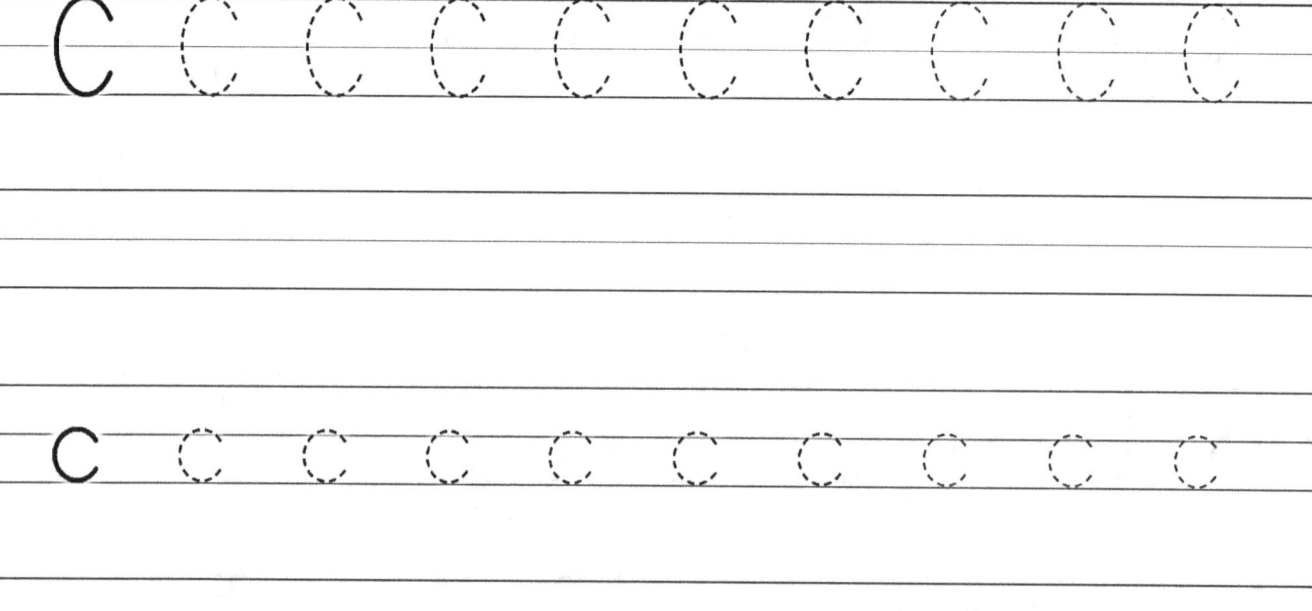

Trace the letters with a pencil. Then practice writing the letters on the lines

C C C C C C C C C C

c c c c c c c c c c

Letter Tracing
Alphabet Writing Practice

Name:

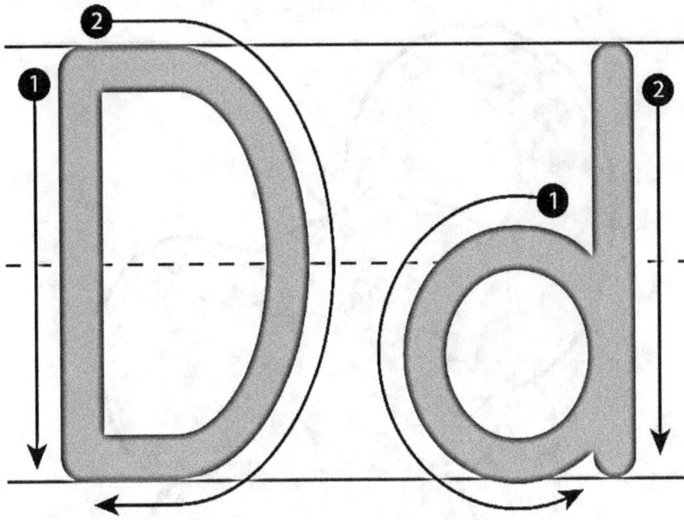

Dd

D is for
Deer

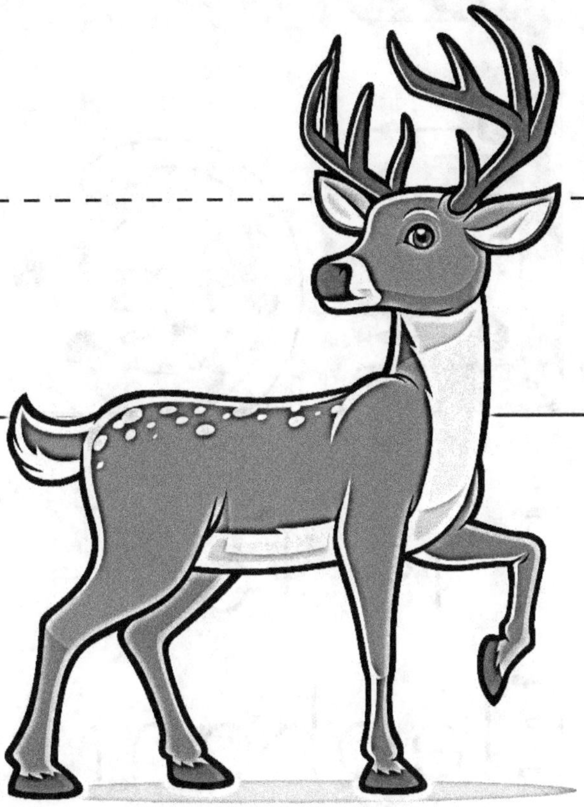

Trace the letters with a pencil. Then practice writing the letters on the lines

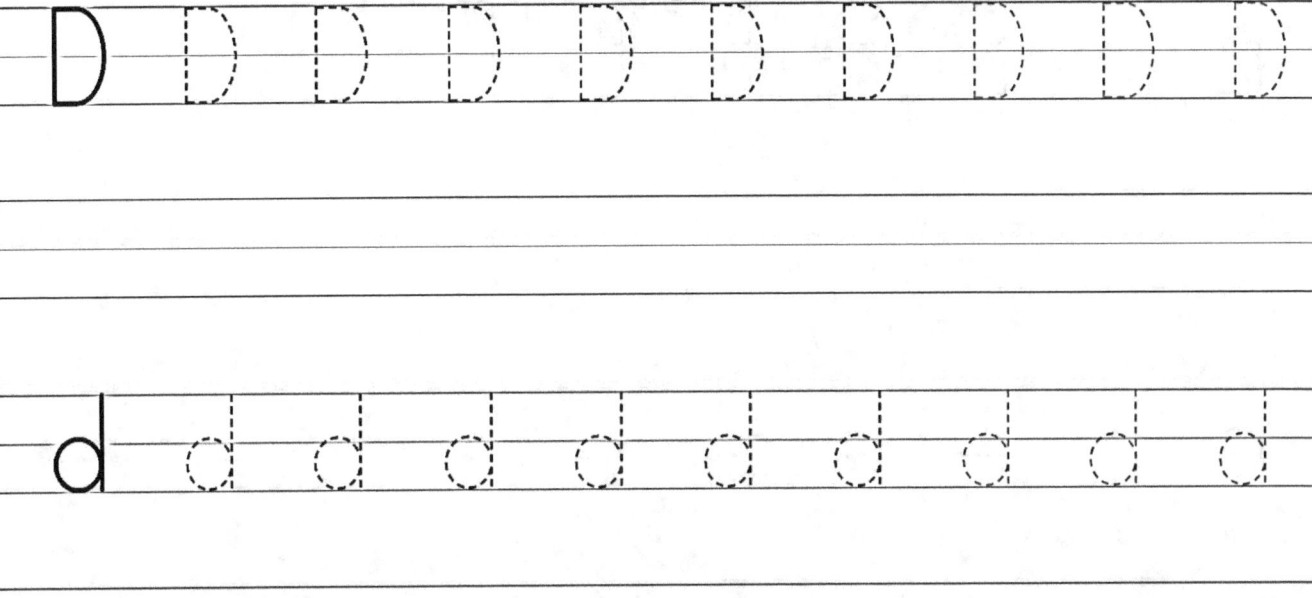

Letter Tracing
Alphabet Writing Practice

Name:

E is for
Elephant

Trace the letters with a pencil. Then practice writing the letters on the lines

Letter Tracing
Alphabet Writing Practice

Name:

F is for Frog

Trace the letters with a pencil. Then practice writing the letters on the lines

Letter Tracing
Alphabet Writing Practice

Name: _____

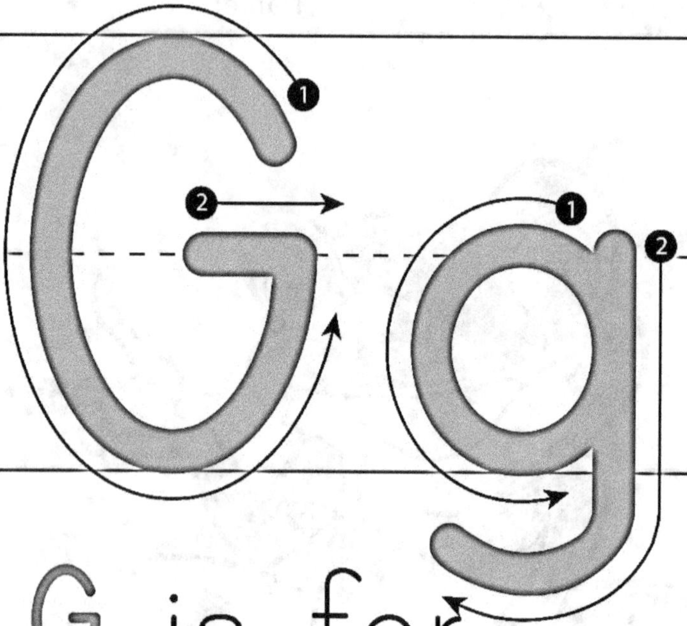

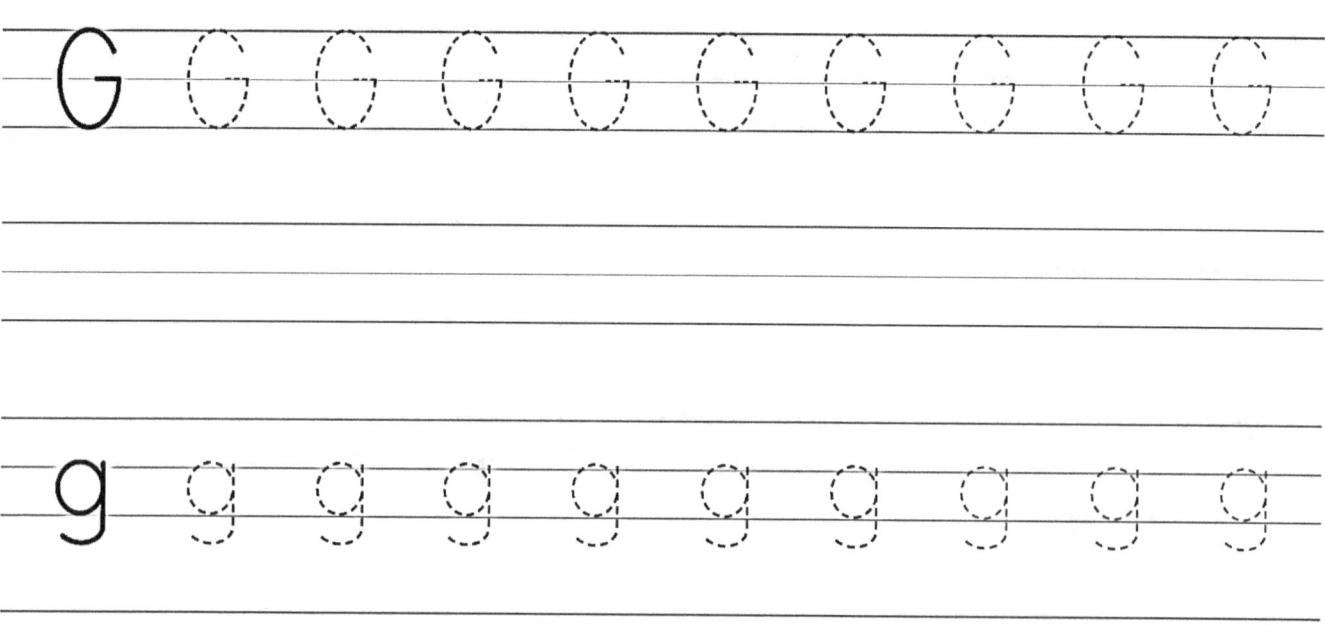

G is for Goldfish

Trace the letters with a pencil. Then practice writing the letters on the lines

G G G G G G G G G G

g g g g g g g g g g

Letter Tracing
Alphabet Writing Practice

Name: _____

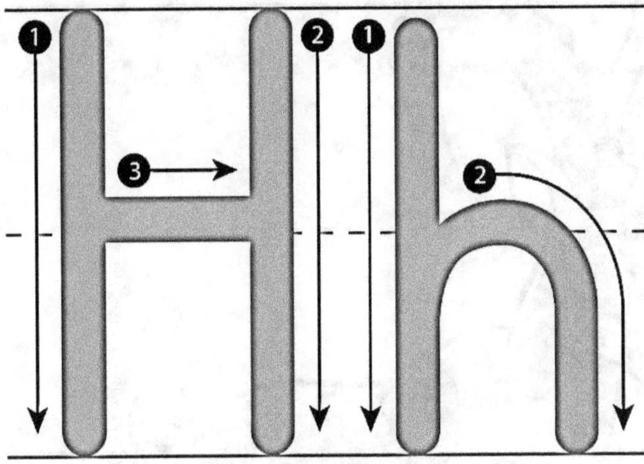

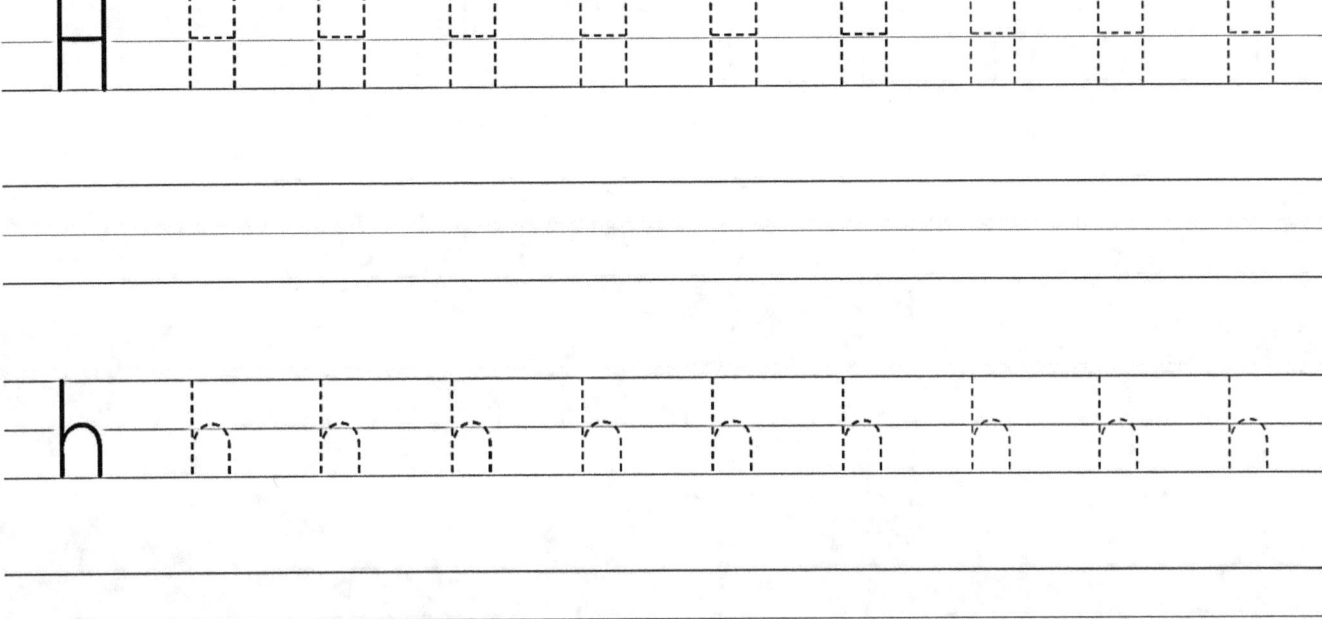

H is for Horse

Trace the letters with a pencil. Then practice writing the letters on the lines

Letter Tracing
Alphabet Writing Practice

Name: _____

I is for
Ibex

Trace the letters with a pencil. Then practice writing the letters on the lines

Letter Tracing
Alphabet Writing Practice

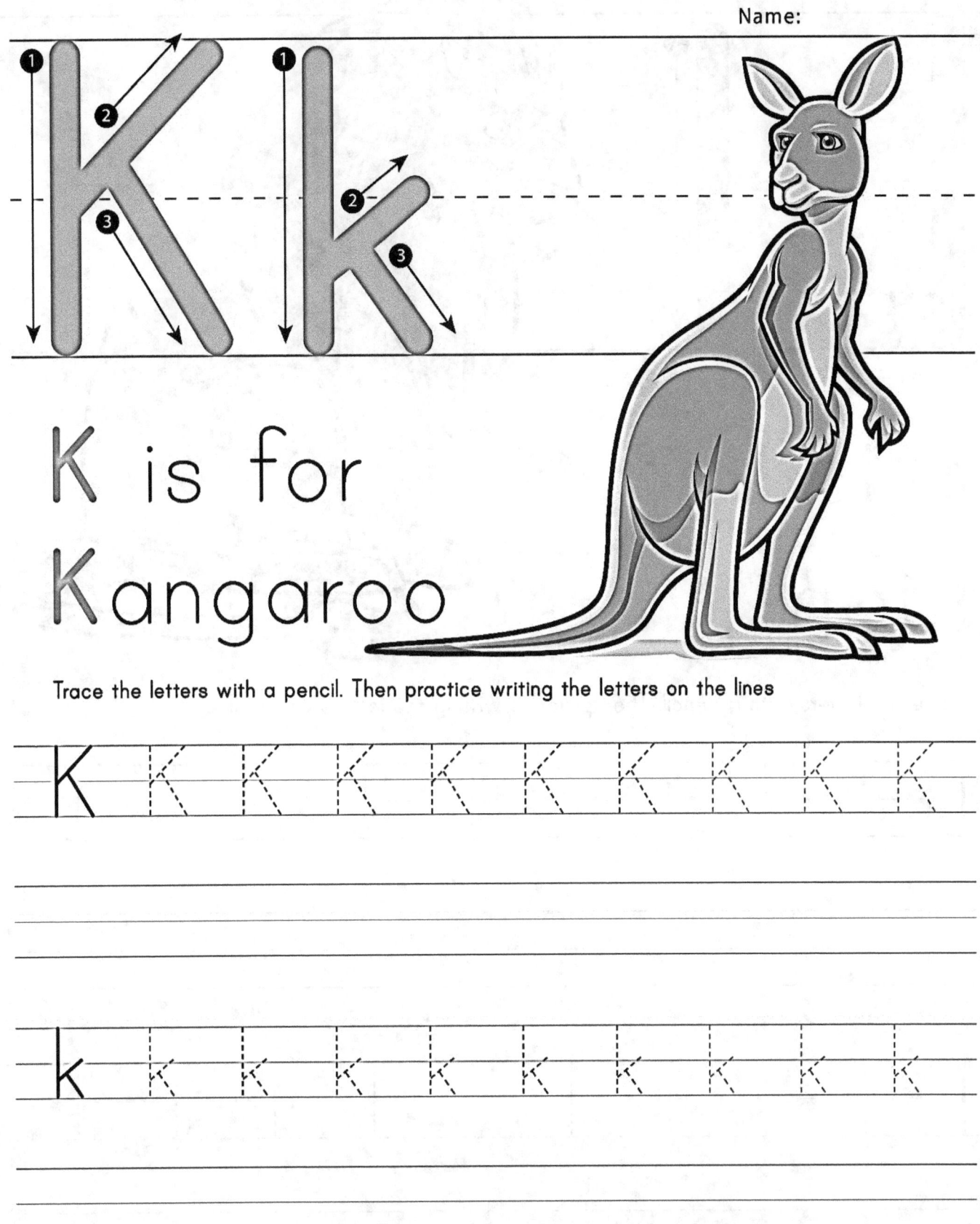

Name:

K is for
Kangaroo

Trace the letters with a pencil. Then practice writing the letters on the lines

Letter Tracing
Alphabet Writing Practice

Name: _____

L is for Lion

Trace the letters with a pencil. Then practice writing the letters on the lines

Letter Tracing
Alphabet Writing Practice

Name: _____

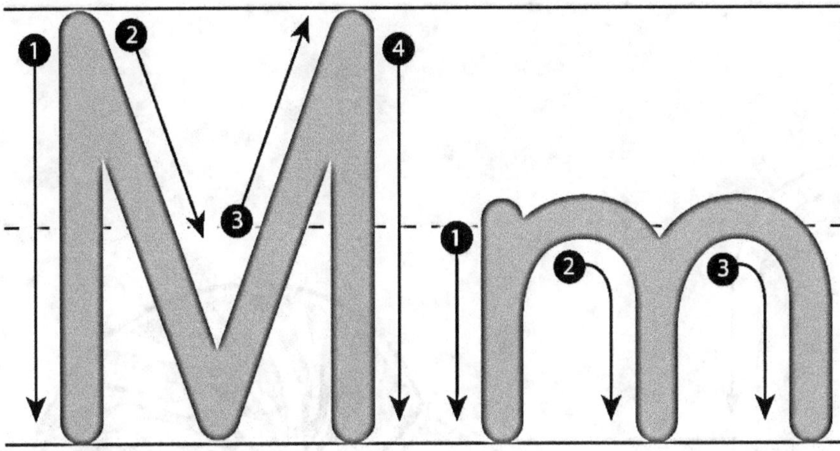

Mm

M is for
Mouse

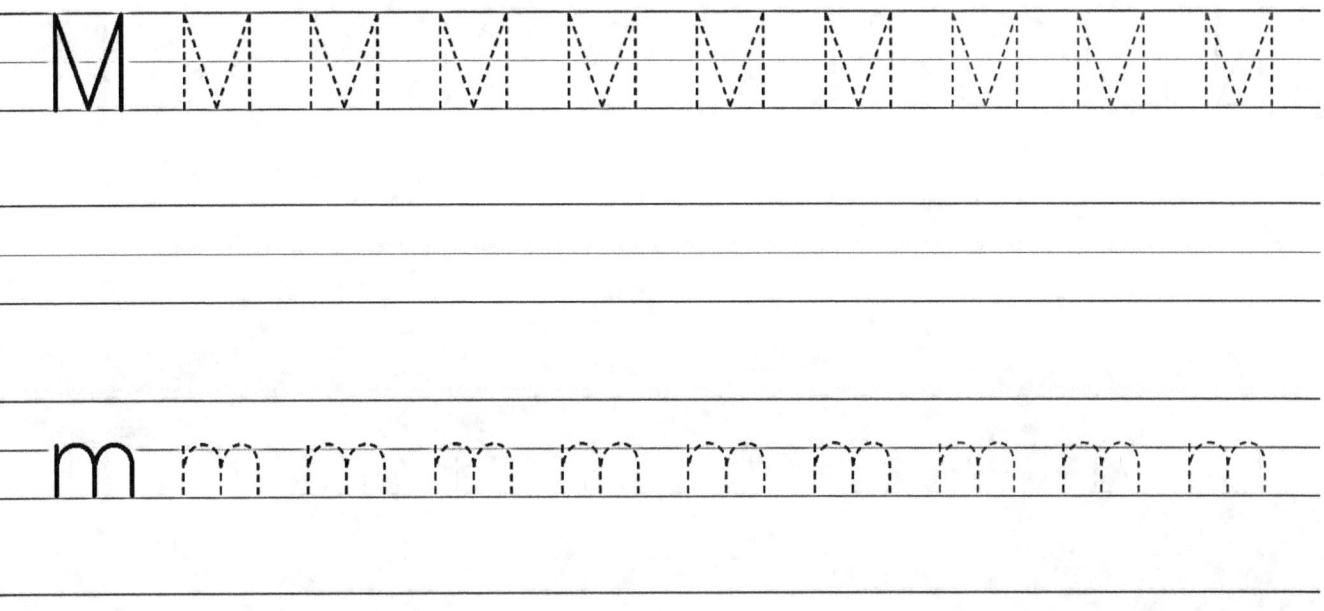

Trace the letters with a pencil. Then practice writing the letters on the lines

M M M M M M M M M M

m m m m m m m m m m

Letter Tracing
Alphabet Writing Practice

Name:

N n

N is for Nautilus

Trace the letters with a pencil. Then practice writing the letters on the lines

N N N N N N N N N N

n n n n n n n n n n

Letter Tracing
Alphabet Writing Practice

Name:

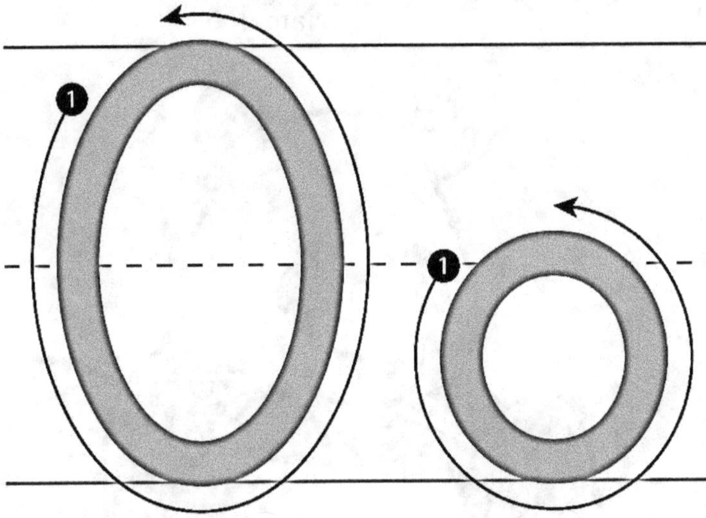

O is for
Owl

Trace the letters with a pencil. Then practice writing the letters on the lines

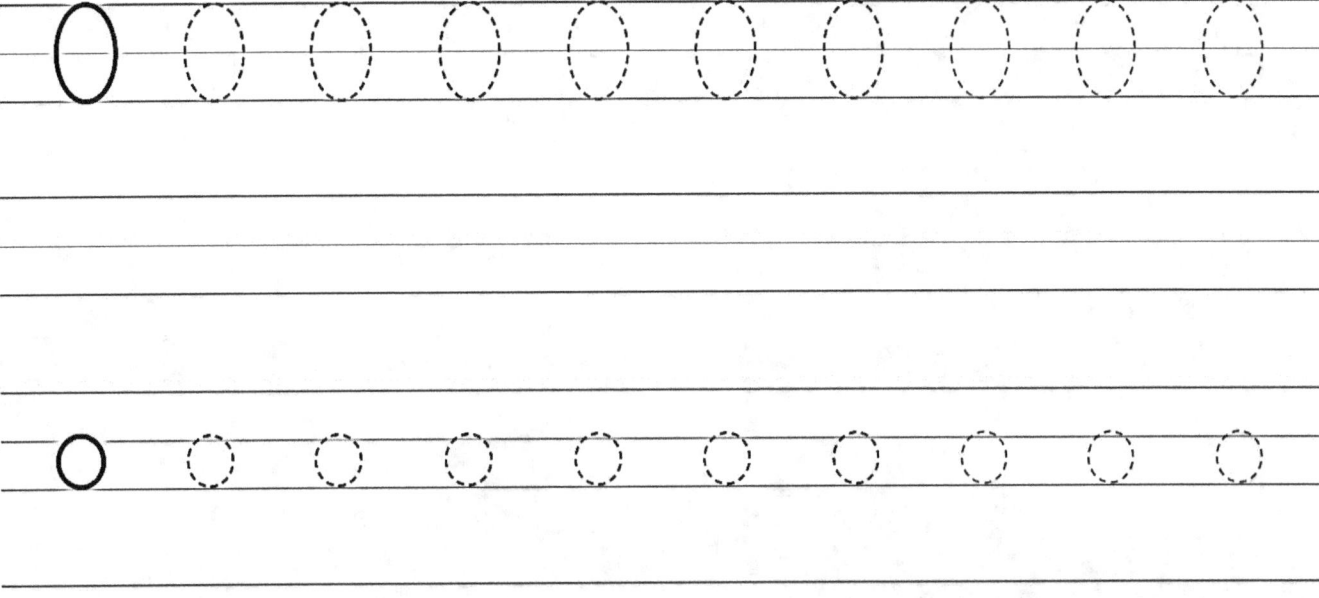

Letter Tracing
Alphabet Writing Practice

Name: _____

P is for Panda

Trace the letters with a pencil. Then practice writing the letters on the lines

P p p p p p p p p p p

_____
_____
_____

p p p p p p p p p p p

_____
_____
_____

Letter Tracing
Alphabet Writing Practice

Name:

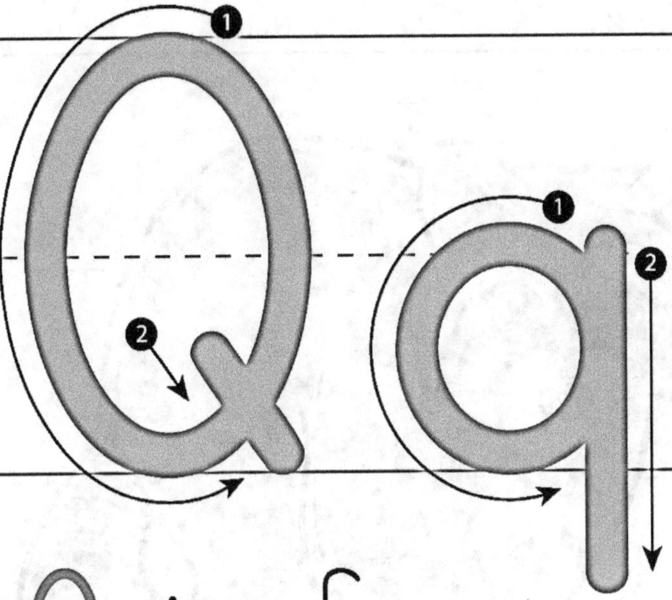

Q q

Q is for Quail

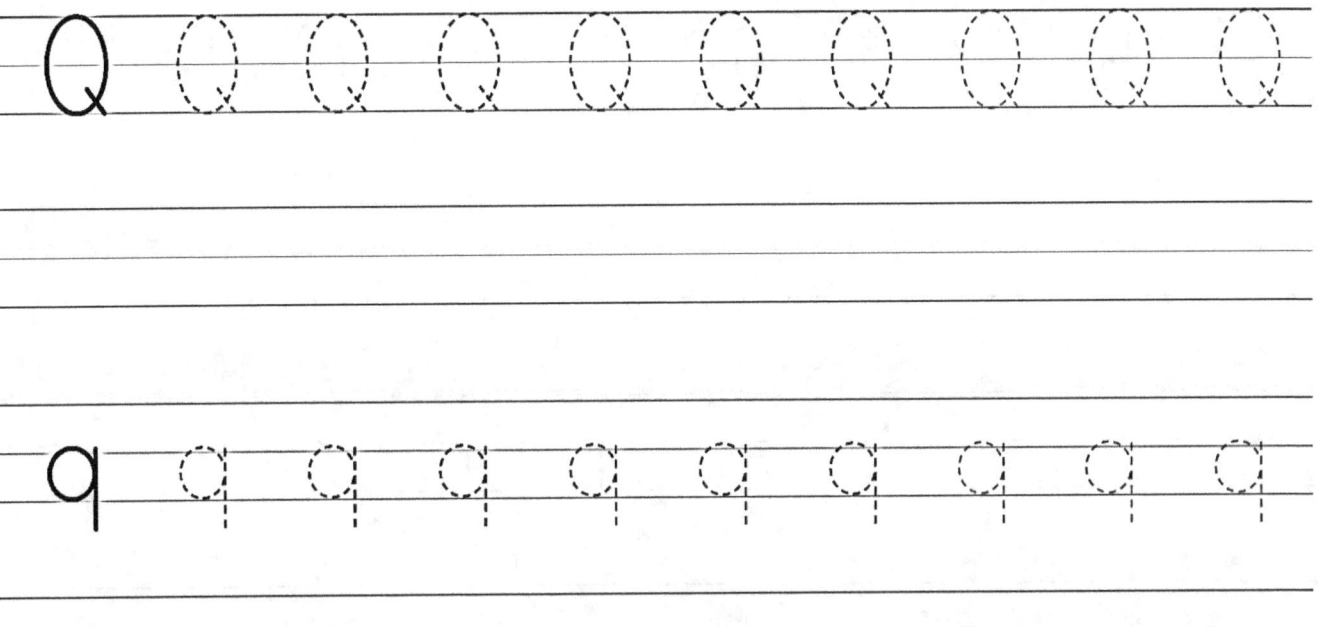

Trace the letters with a pencil. Then practice writing the letters on the lines

Q Q Q Q Q Q Q Q Q Q Q

q q q q q q q q q q q

Letter Tracing
Alphabet Writing Practice

Name: _____

R is for Raccoon

Trace the letters with a pencil. Then practice writing the letters on the lines

R R R R R R R R R

r r r r r r r r r r

Letter Tracing
Alphabet Writing Practice

Name: _____

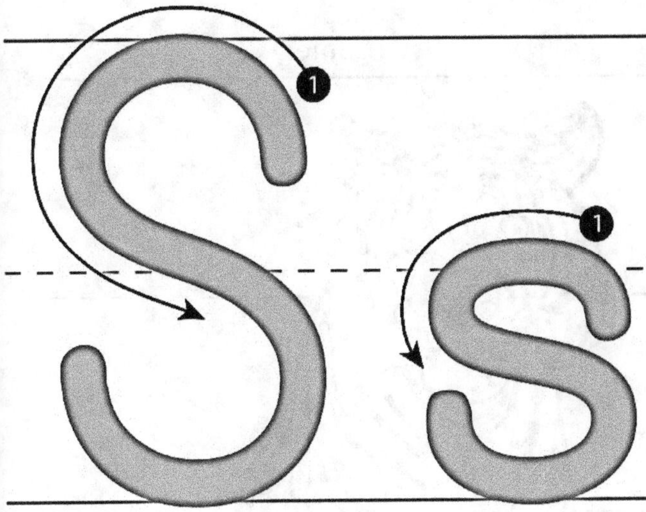

S s

S is for
Squirrel

Trace the letters with a pencil. Then practice writing the letters on the lines

S S S S S S S S S S

s s s s s s s s s s

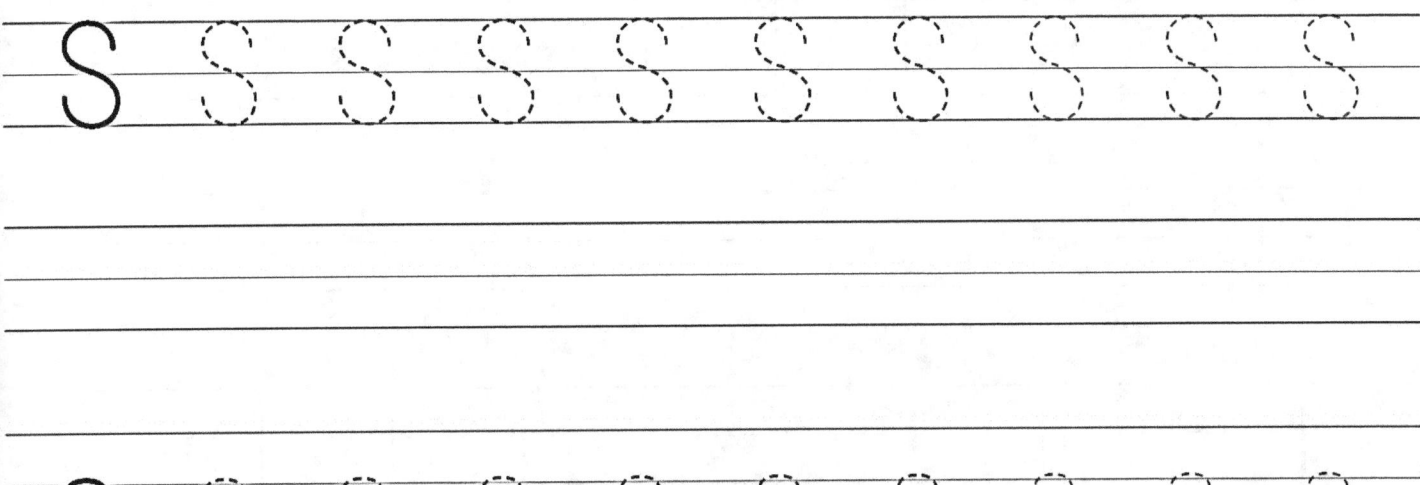

Letter Tracing
Alphabet Writing Practice

Name:

T is for
Tiger

Trace the letters with a pencil. Then practice writing the letters on the lines

Letter Tracing
Alphabet Writing Practice

Name: _____

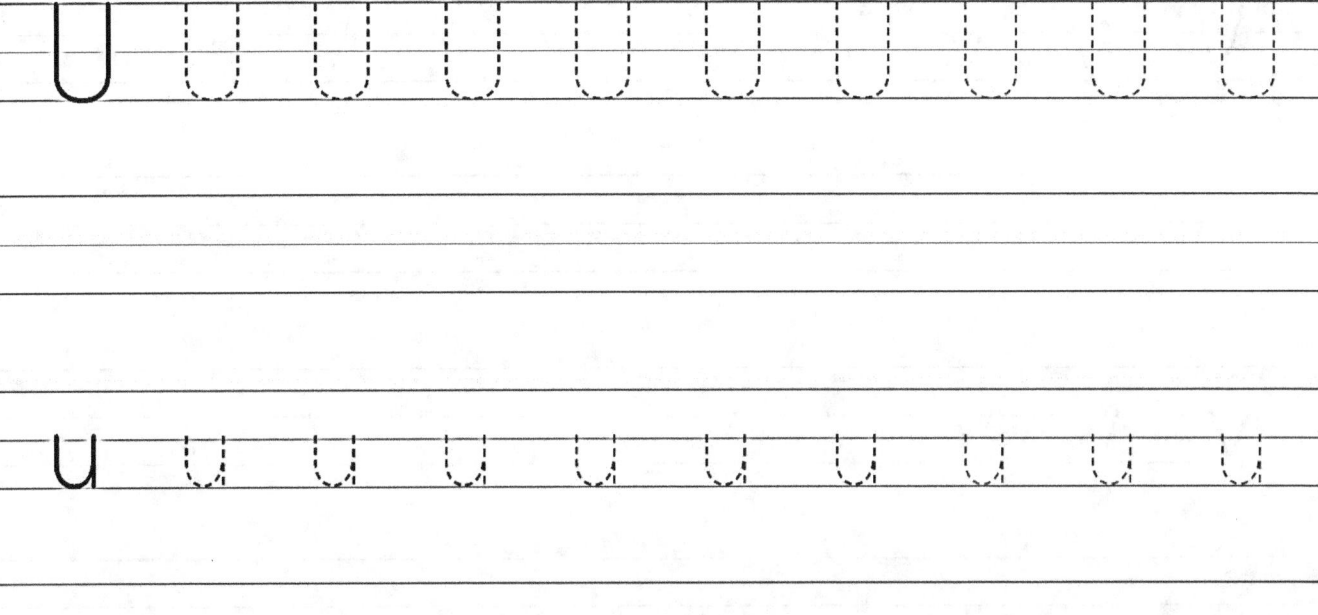

U is for
Umbrellabird

Trace the letters with a pencil. Then practice writing the letters on the lines

U U U U U U U U U U U

u u u u u u u u u u u

Letter Tracing
Alphabet Writing Practice

Name:

V is for
Vulture

Trace the letters with a pencil. Then practice writing the letters on the lines

Letter Tracing
Alphabet Writing Practice

Name: _____

W is for Wolf

Trace the letters with a pencil. Then practice writing the letters on the lines

Letter Tracing
Alphabet Writing Practice

Name: _____

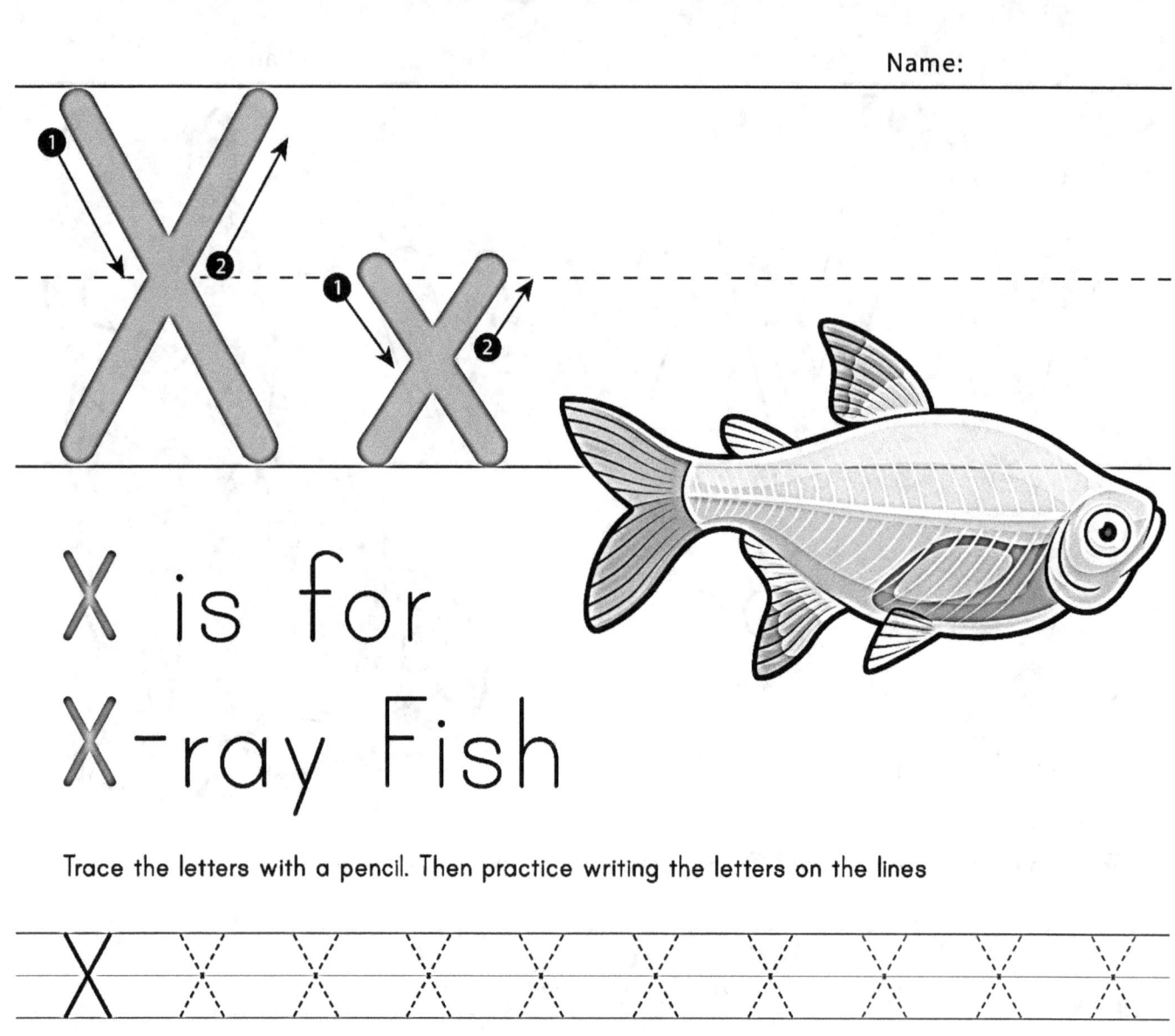

X is for X-ray Fish

Trace the letters with a pencil. Then practice writing the letters on the lines

Letter Tracing
Alphabet Writing Practice

Name:

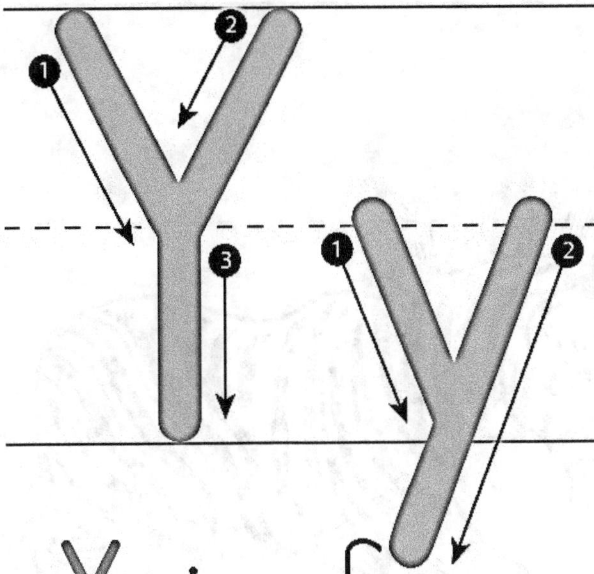

Y is for
Yak

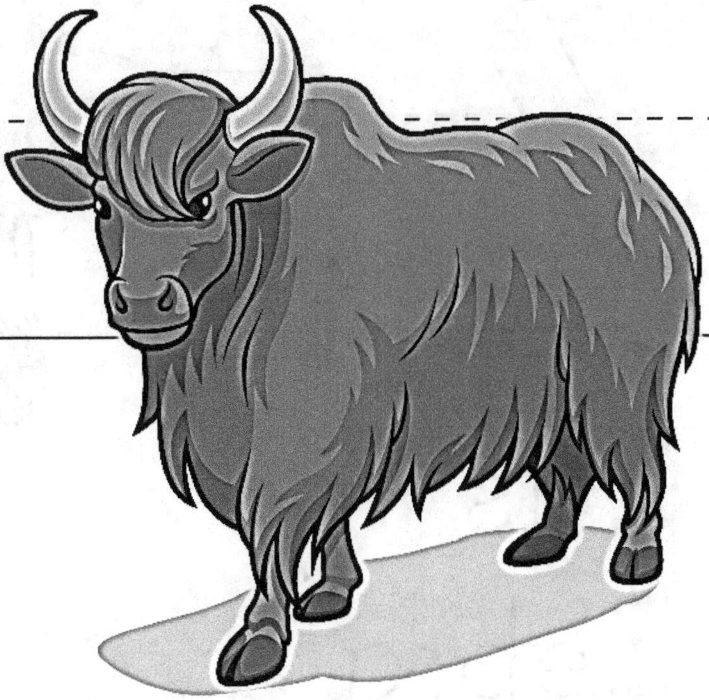

Trace the letters with a pencil. Then practice writing the letters on the lines

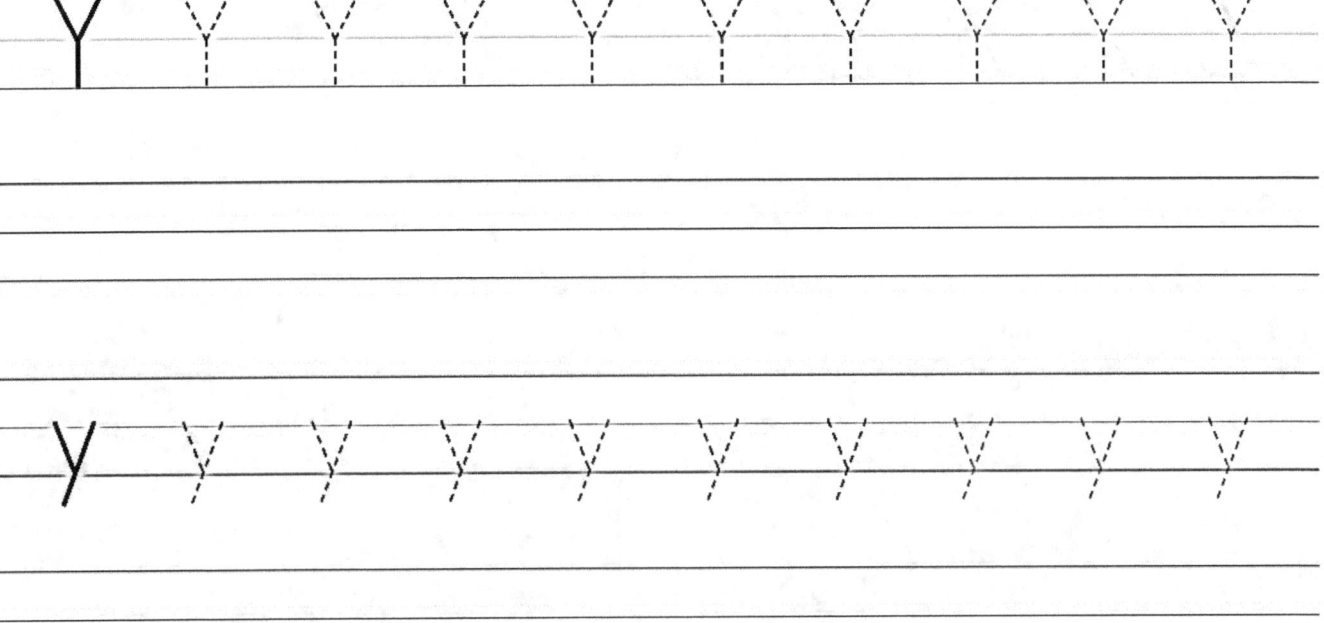

Letter Tracing
Alphabet Writing Practice

Name: _____

Z is for Zebra

Trace the letters with a pencil. Then practice writing the letters on the lines

Math
Addition & Subtraction

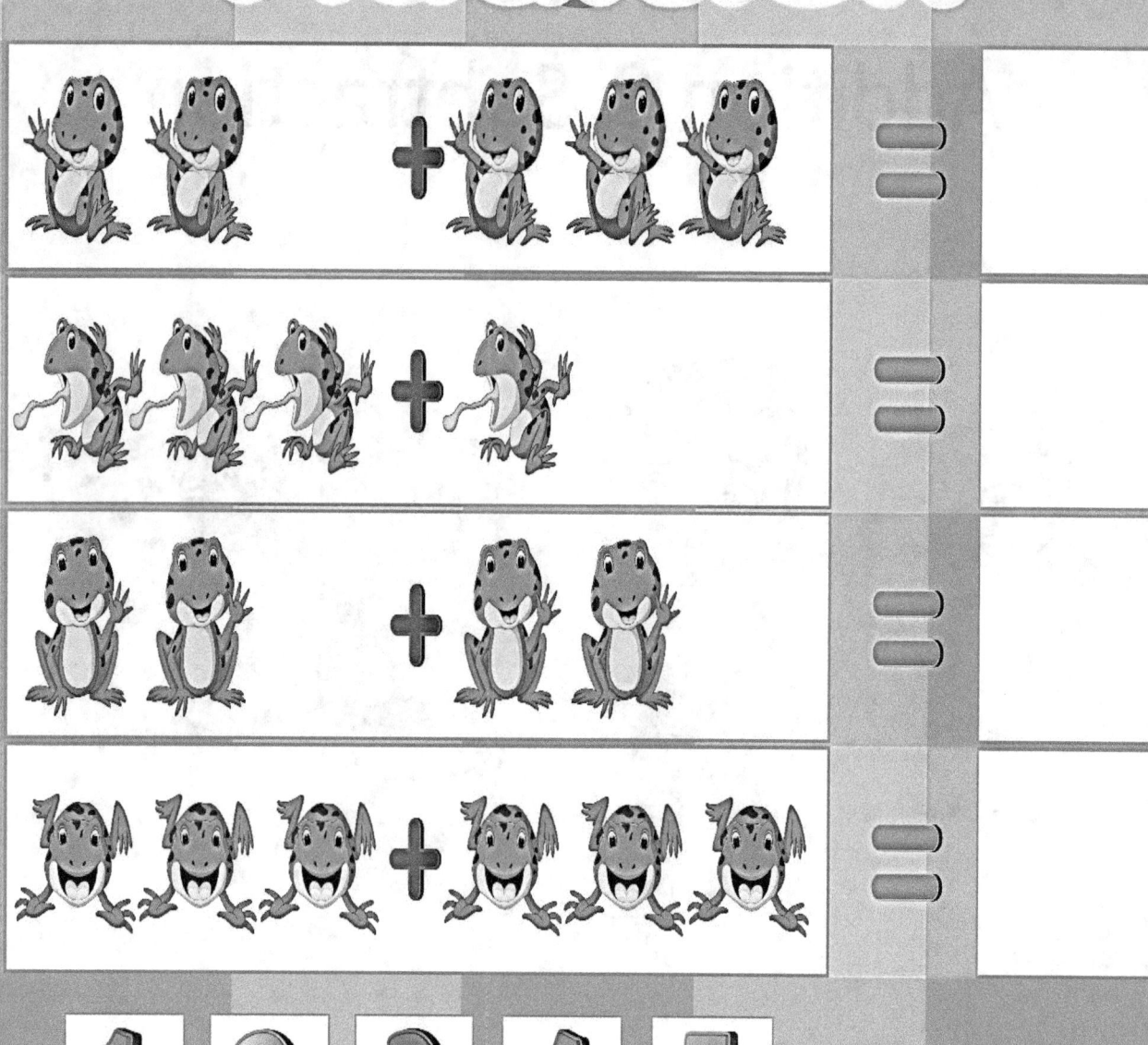

Addition number

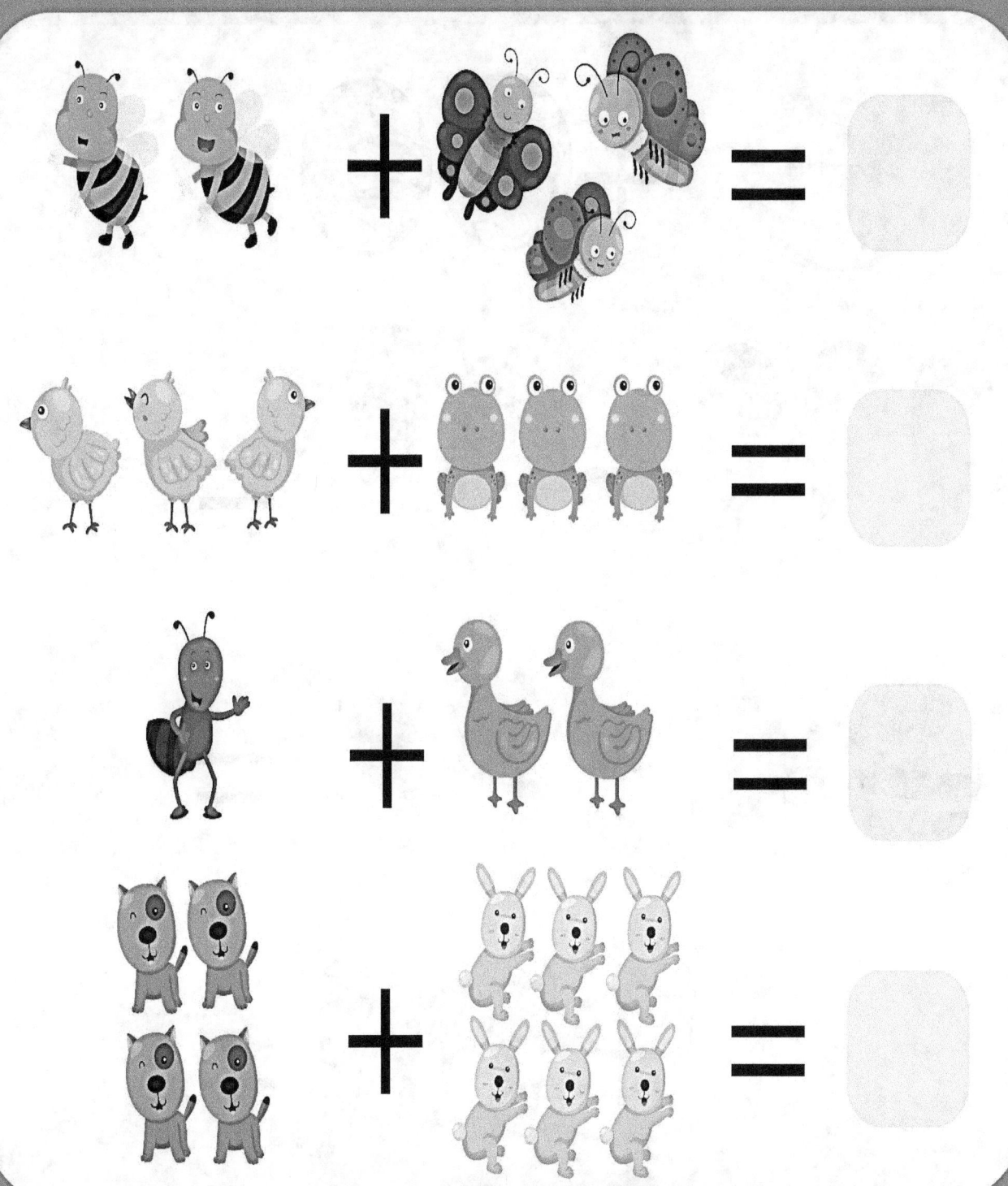

Name .... Class ....

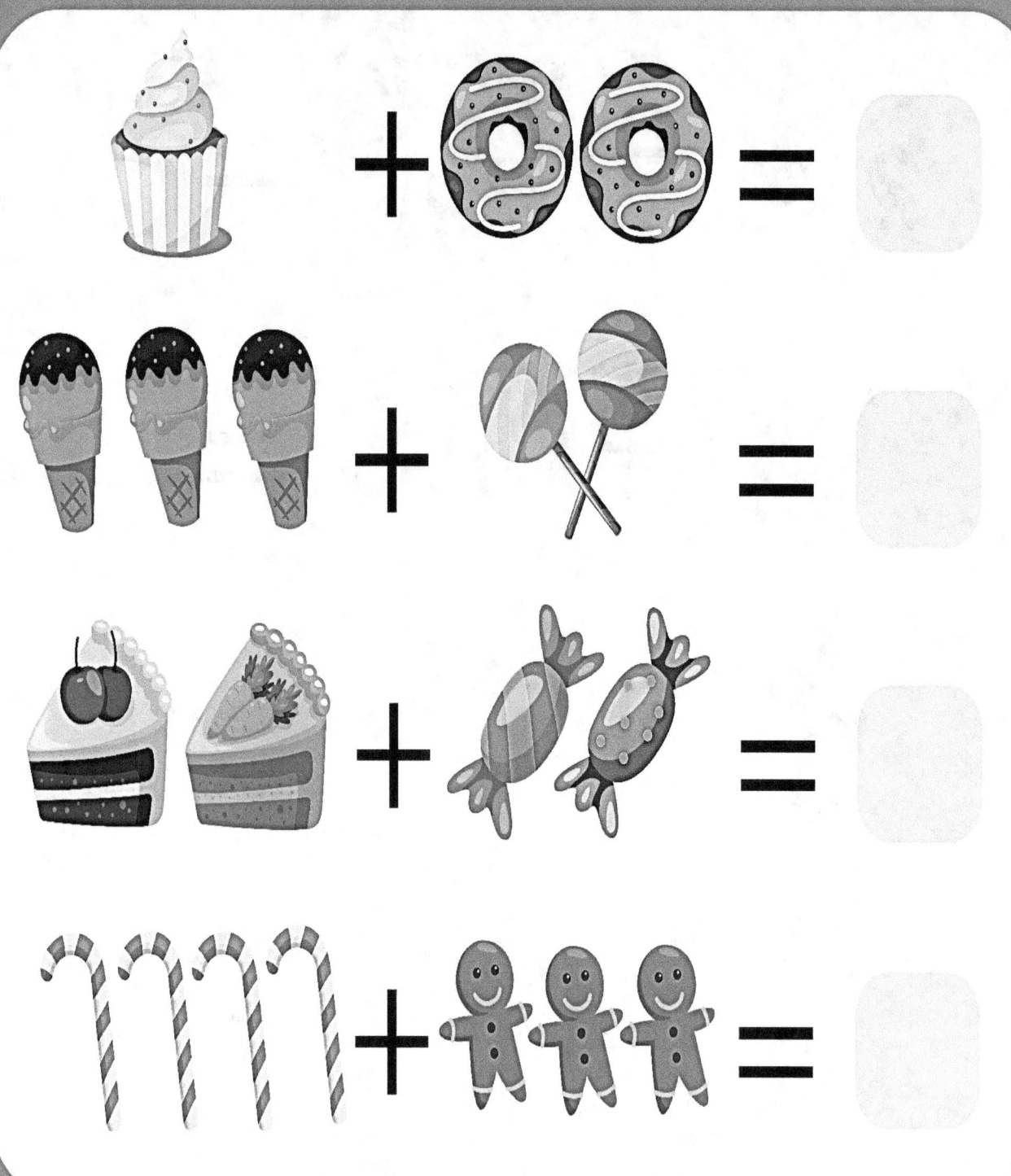

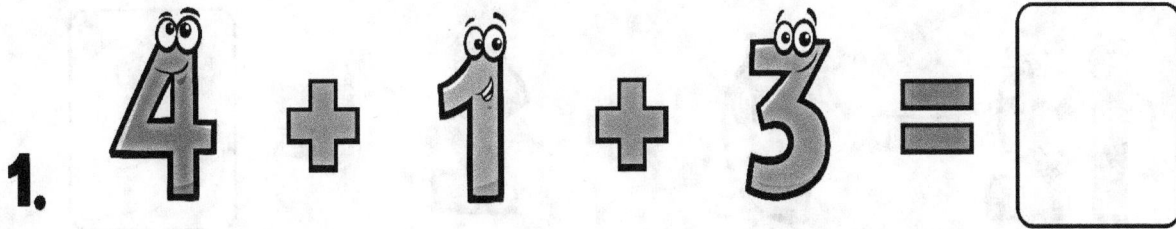

1. 4 + 1 + 3 = ☐

2. 9 + 3 − ☐ = 10

3. 8 − 6 − 1 = ☐

4. 7 − ☐ + 5 = 6

5. 5 − 2 + ☐ = 8

1. $12 - 9 + 2 = \boxed{}$

2. $3 + \boxed{} + 4 = 8$

3. $\boxed{} - 5 - 5 = 10$

4. $7 - 6 + \boxed{} = 3$

5. $4 - \boxed{} + 8 = 9$

1. ☐ + 4 − 6 = 10

2. 2 + ☐ + 7 = 17

3. 8 + 5 − 6 = ☐

4. 9 − ☐ + 6 = 12

5. 14 − 5 − ☐ = 7

How many?

5

Subtraction within 10

a. 1
  - 0
  ___

b. 5
  - 4
  ___

c. 8
  - 8
  ___

d. 8
  - 0
  ___

e. 9
  - 1
  ___

f. 4
  - 2
  ___

g. 4
  - 4
  ___

h. 9
  - 5
  ___

i. 8
  - 7
  ___

Additon to 50

a. 20
 + 10
 ———

b. 23
 + 15
 ———

c. 8
 + 10
 ———

d. 6
 + 35
 ———

e. 19
 + 21
 ———

f. 31
 + 4
 ———

g. 18
 + 13
 ———

h. 26
 + 6
 ———

i. 11
 + 15
 ———

Subtraction within 20

a. 32
  − 28
  ———

b. 17
  − 14
  ———

c. 19
  − 15
  ———

d. 29
  − 13
  ———

e. 30
  − 29
  ———

f. 28
  − 16
  ———

g. 49
  − 29
  ———

h. 35
  − 6
  ———

i. 43
  − 20
  ———

Additon to 50

a. 20 + 10

b. 23 + 15

c. 8 + 10

d. 6 + 35

e. 19 + 21

f. 31 + 4

g. 18 + 13

h. 26 + 6

i. 11 + 15

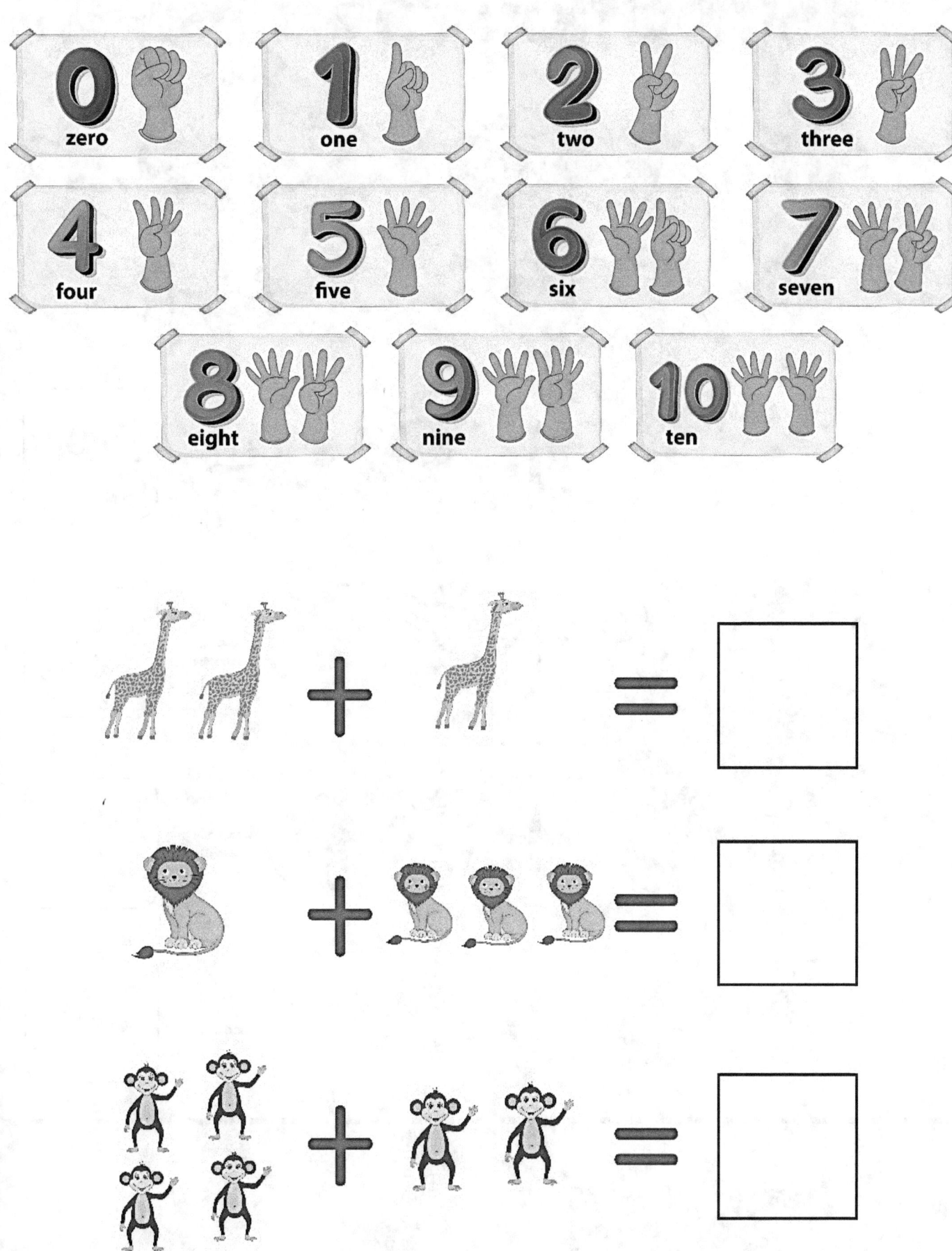

ADDITION FOR KIDS

🍎 + 🍎 = [?]

🍎🍎🍎 + 🍎🍎 = [?]

🍎🍎 + 🍎🍎 = [?]

🍎🍎 + 🍎 = [?]

- -

[2] [4] [5] [3]

ADDITION FOR KIDS

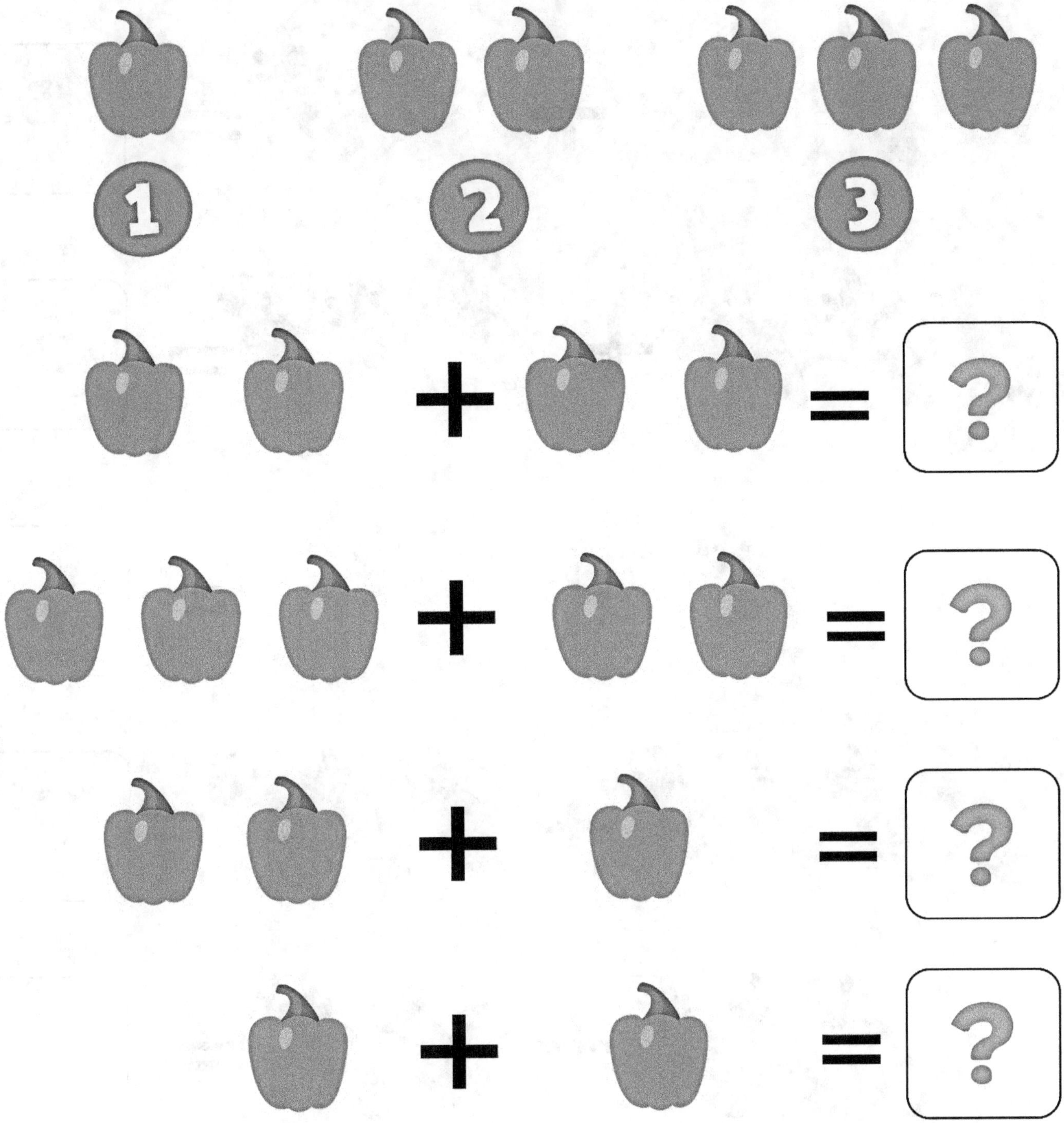

🫑   🫑🫑   🫑🫑🫑
①   ②   ③

🫑🫑 + 🫑🫑 = [?]

🫑🫑🫑 + 🫑🫑 = [?]

🫑🫑 + 🫑 = [?]

🫑 + 🫑 = [?]

- -

ADDITION FOR KIDS

🥕🥕 + 🥕🥕 = [?]

🥕🥕🥕 + 🥕🥕🥕 = [?]

🥕 + 🥕 = [?]

🥕🥕🥕🥕 + 🥕🥕🥕 = [?]

🥕🥕🥕 + 🥕🥕 = [?]

- -

[4] [5] [6] [2] [7]

SUBTRACTION FOR KIDS

4 − 1 = 3

4 − 2 = ?

5 − 3 = ?

5 − 4 = ?

3 − 2 = ?

4 − 3 = ?

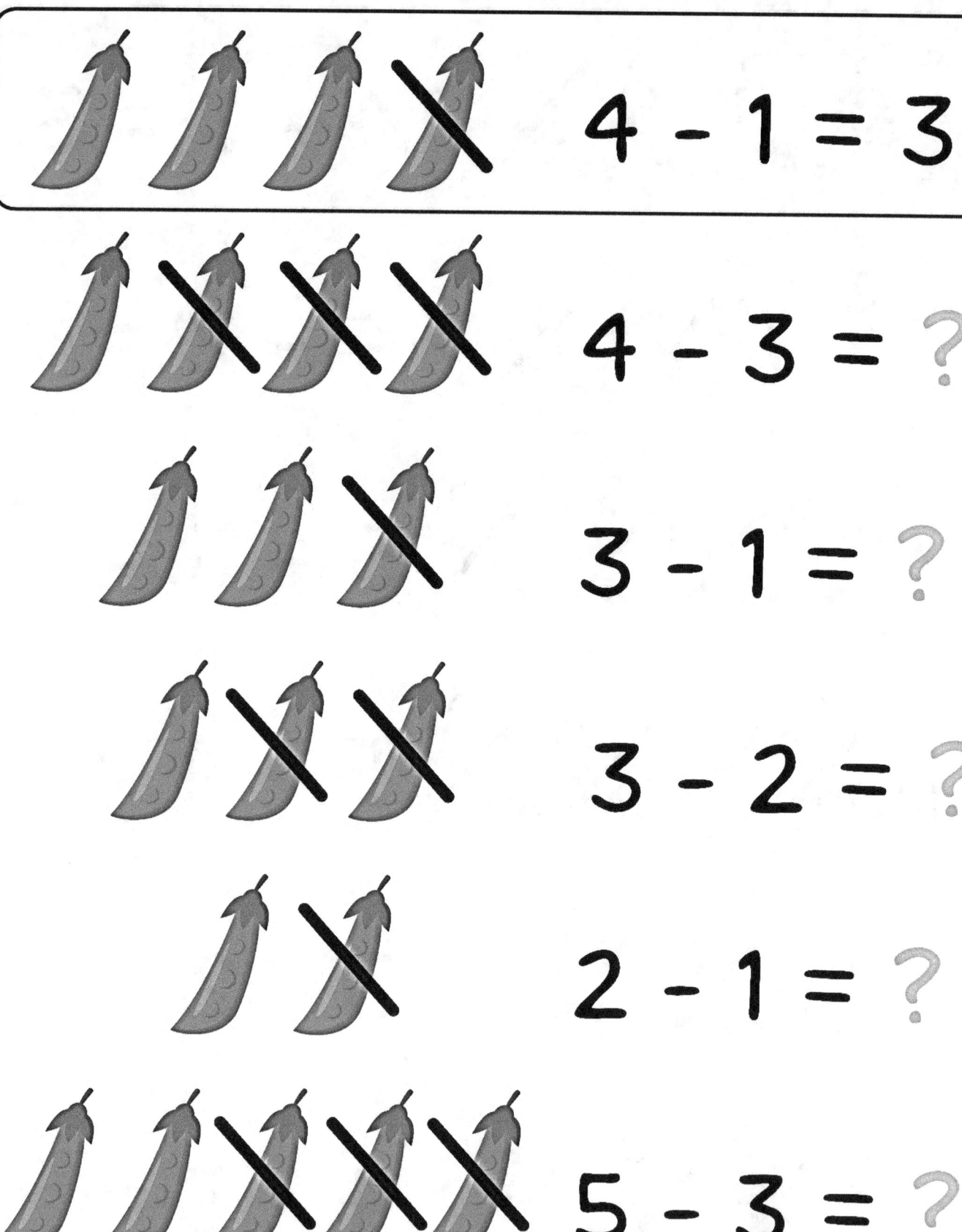

SUBTRACTION FOR KIDS

4 − 1 = 3

4 − 2 = ?

5 − 3 = ?

5 − 2 = ?

3 − 2 = ?

3 − 1 = ?

SUBTRACTION FOR KIDS

 3 - 2 = 2

 3 - 1 = ?

 3 - 3 = ?

 4 - 2 = ?

 4 - 3 = ?

 4 - 1 = ?

SUBTRACTION FOR KIDS

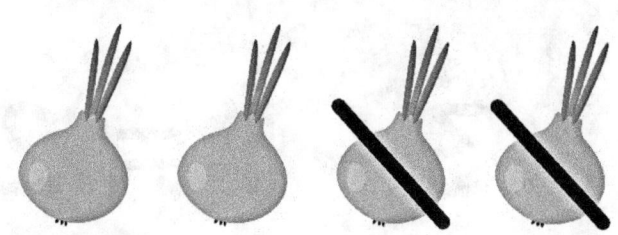 4 - 2 = 2

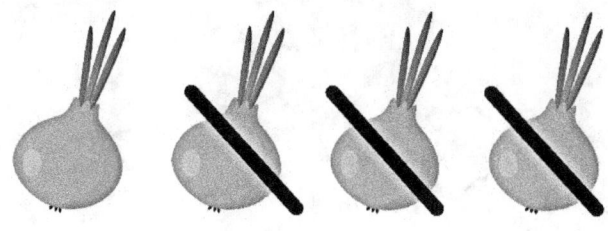 4 - 3 = ?

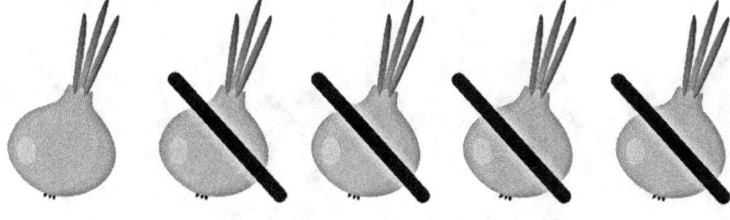 5 - 4 = ?

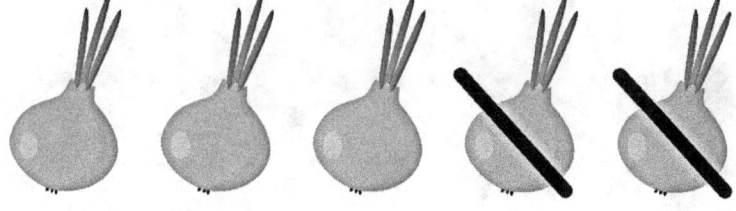 5 - 2 = ?

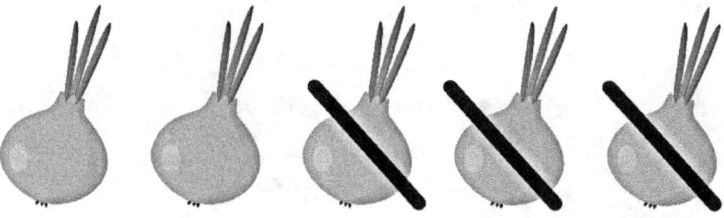 5 - 3 = ?

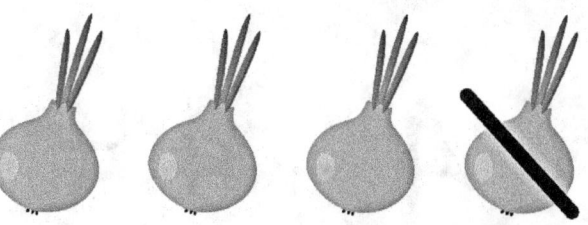 4 - 1 = ?

SUBTRACTION FOR KIDS

     3 - 1 = 2

     3 - 2 = ?

      4 - 3 = ?

   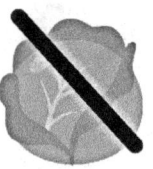   4 - 1 = ?

     3 - 2 = ?

      4 - 4 = ?

Circle the right answer

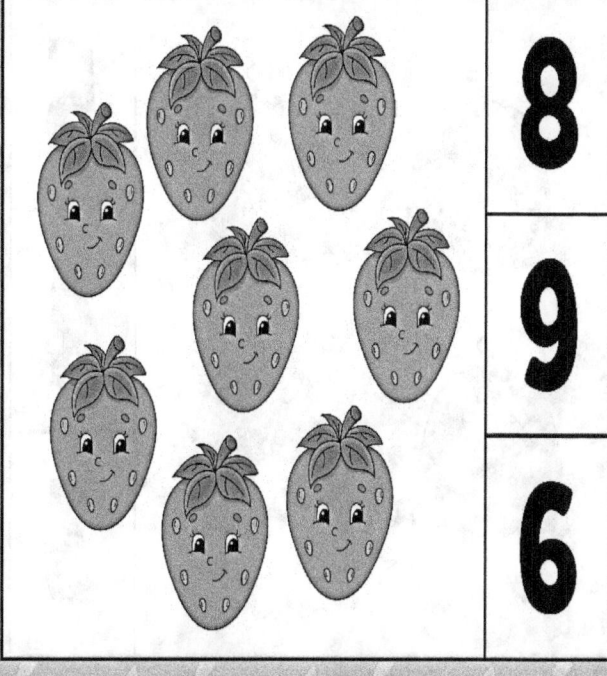
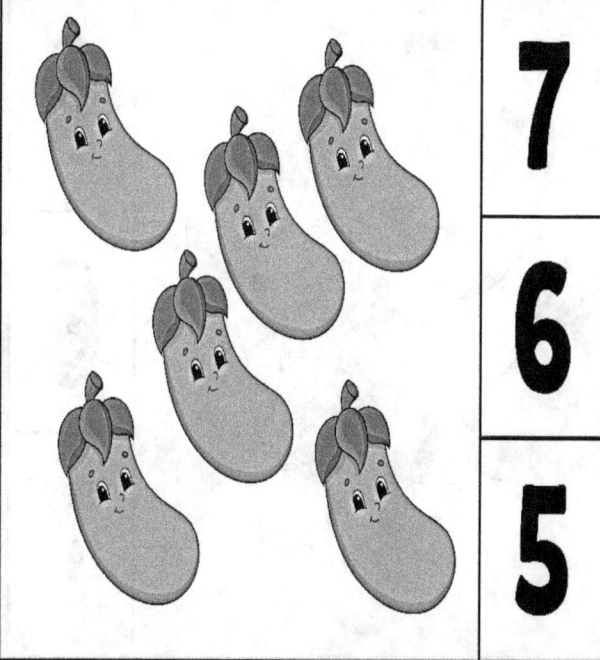

Circle the right answer

Math game for kids

12 + 8 =

13 − 3 =

15 + 2 =

17 20 10

Math game for kids

6 + 3 =

14 − 6 =

5 + 2 =

8 9 7

Math game for kids

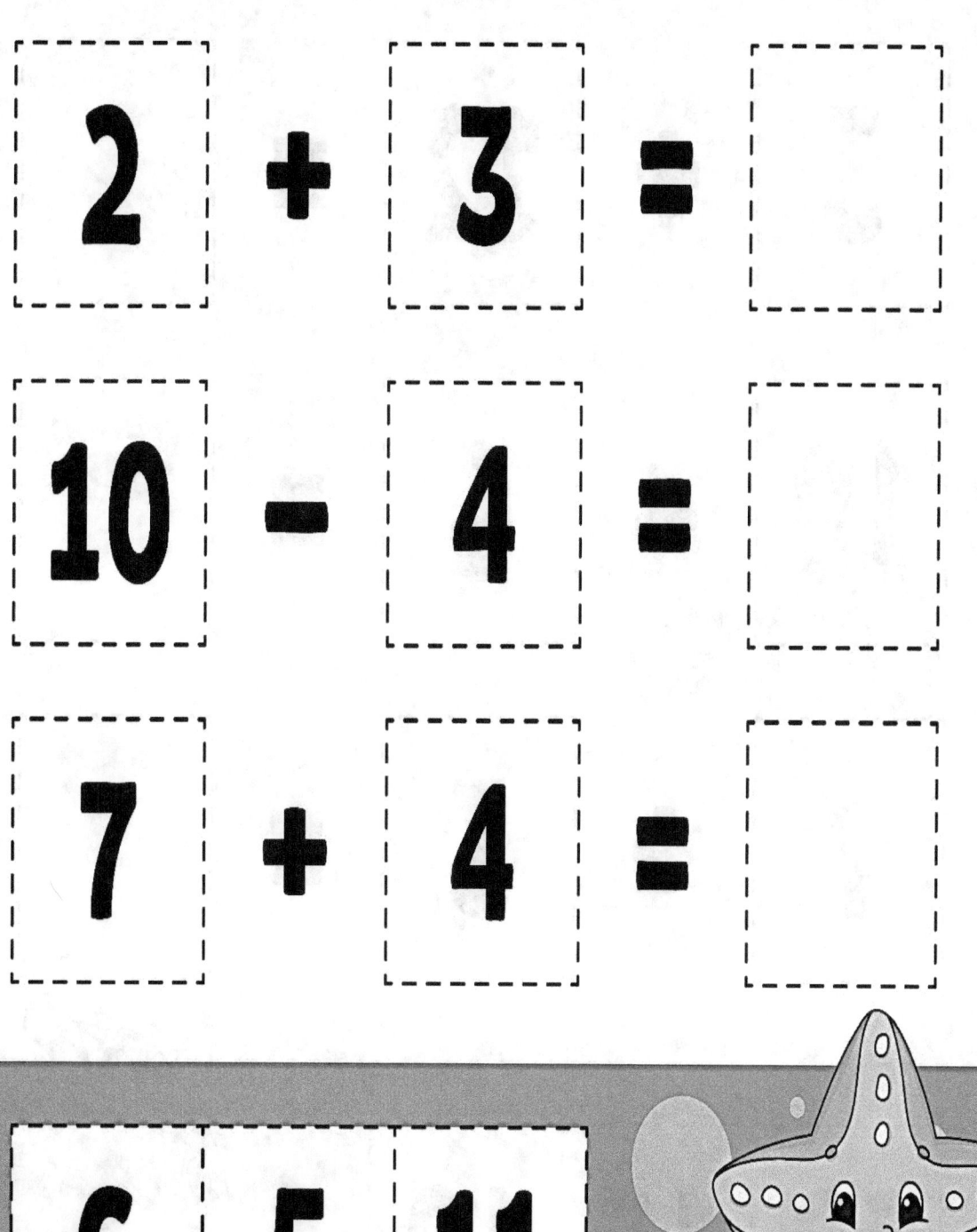

Math game for kids

3 + 12 =

18 − 6 =

7 + 3 =

12　15　10

Math game for kids

15 + 3 =

5 − 2 =

7 + 4 =

11 3 18

Math game for kids

14 + 3 =

19 − 9 =

5 + 13 =

10 17 18

Math game for kids

7 + 3 =

6 − 4 =

2 + 6 =

8   2   10

Math game for kids

18 + 2 =

17 − 7 =

15 + 2 =

17 20 10

HOW MANY?

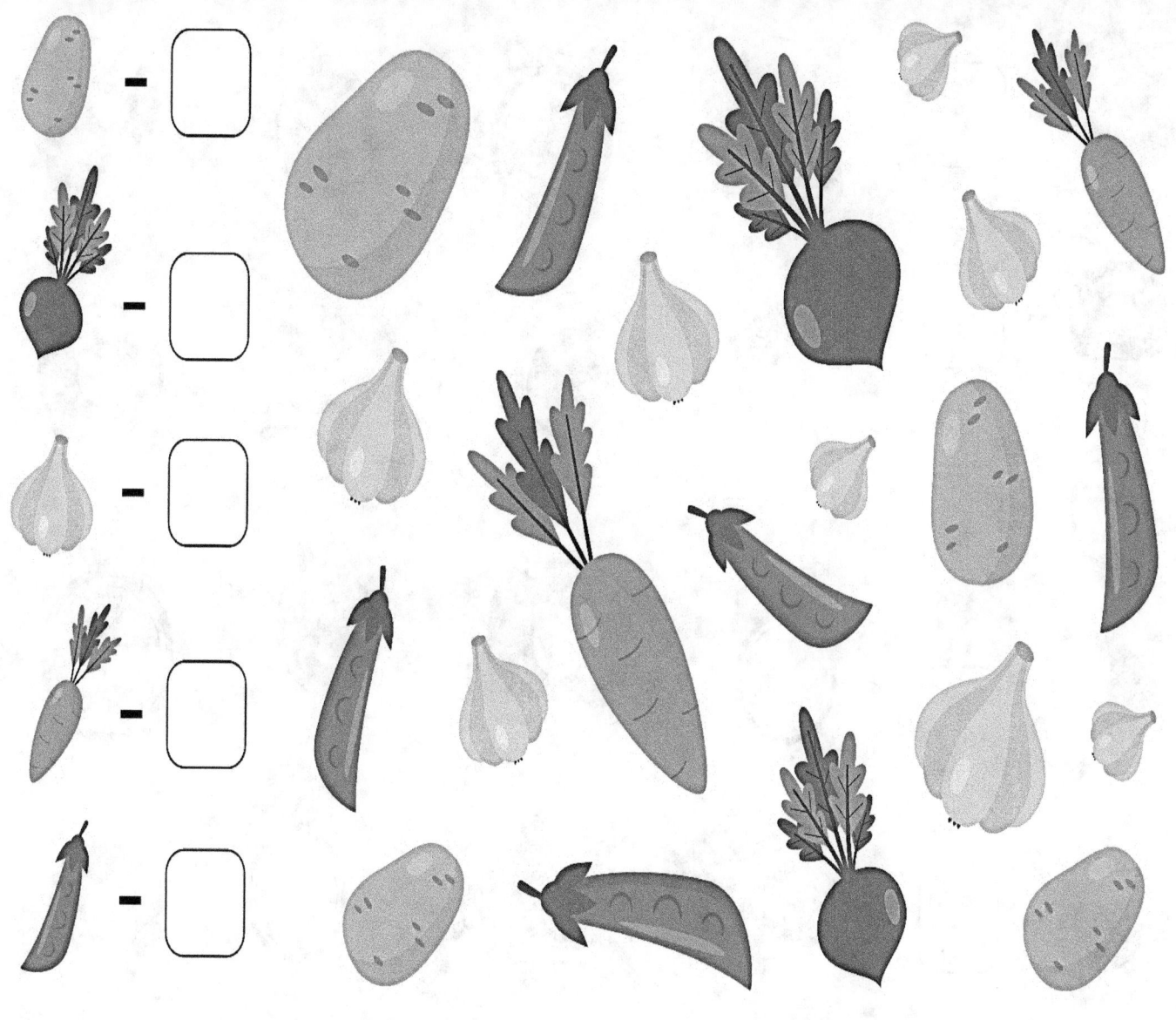

HOW MANY?

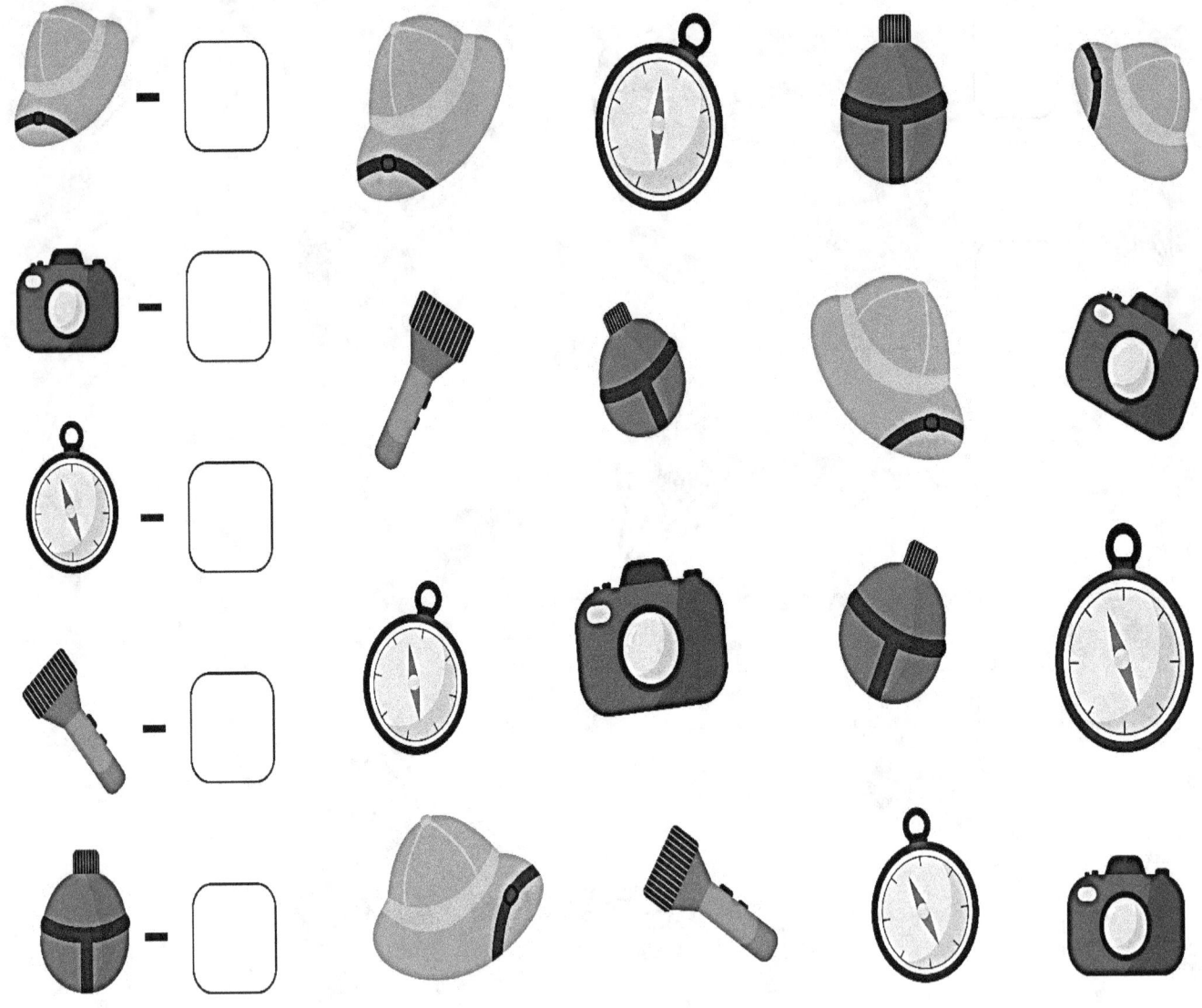

HOW MANY?

Counting Game

7 + 4 = ☐

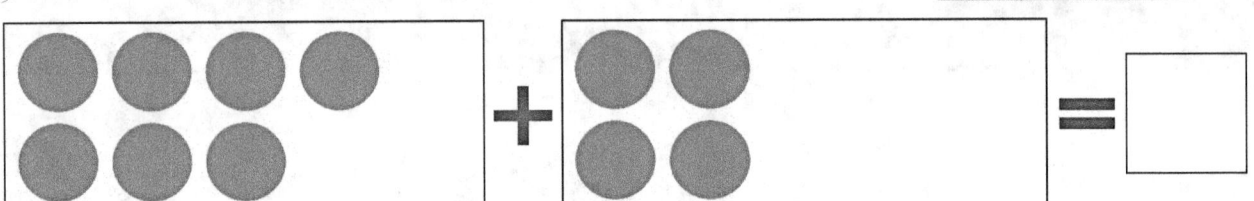

8 − 7 = ☐

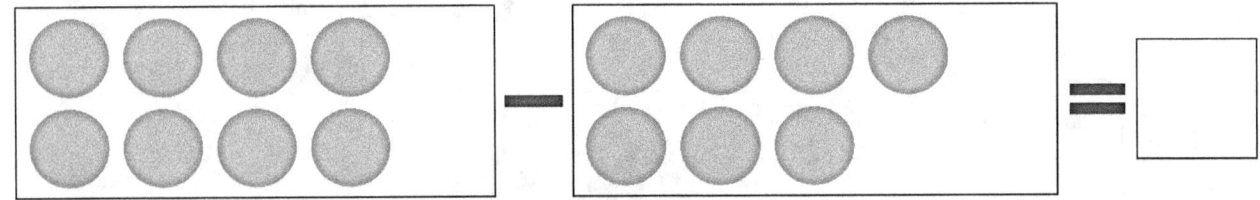

7 + 8 = ☐

3 − 3 = ☐

8 + 9 = ☐

Counting Game

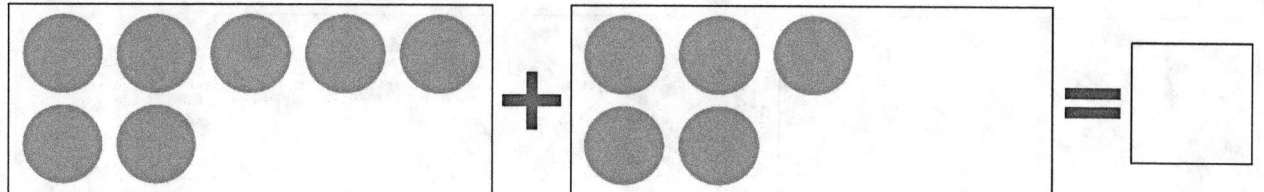

7 + 5 = ☐

6 − 5 = ☐

4 + 6 = ☐

4 − 3 = ☐

4 + 5 = ☐

Counting Game

4 + 8 = ☐

4 − 2 = ☐

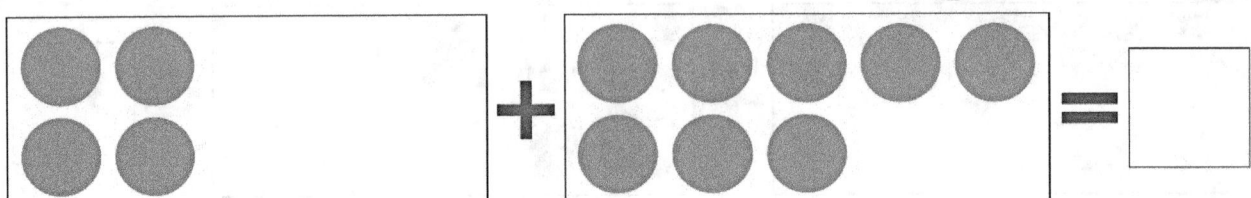

5 − 3 = ☐

7 + 9 = ☐

Counting Game

5 + 6 = ☐

7 − 4 = ☐

1 + 5 = ☐

6 − 3 = ☐

8 + 7 = ☐

Counting Game

8 + 6 = ☐

10 − 8 = ☐

7 + 7 = ☐

8 − 6 = ☐

7 + 7 = ☐

Counting Game

10 + 5 = ☐

4 − 1 = ☐

7 + 3 = ☐

8 − 10 = ☐

7 + 9 = ☐

Addition and Subtraction

2	+	3	=	
4	−	1	=	
3	+	8	=	
6	−	5	=	
5	+	10	=	

5
3
11
1
15

Addition and Subtraction

9	+	1	=	
4	-	3	=	
5	+	4	=	
9	-	5	=	
3	+	10	=	

10

1

3

4

13

Addition and Subtraction

1	+	3	=	
5	-	2	=	
10	+	1	=	
9	-	5	=	
2	+	5	=	

4
3
11
4
7

Addition and Subtraction

1	+	3	=		4
7	−	2	=		5
12	+	3	=		15
18	−	8	=		10
11	+	5	=		16

Addition and Subtraction

2	+	3	=	
4	−	1	=	
10	+	2	=	
7	−	3	=	
9	+	3	=	

5
3
12
4
12

Addition and Subtraction

4 + 7 =	
2 - 1 =	
10 + 3 =	
4 - 2 =	
6 + 6 =	

- 11
- 1
- 13
- 2
- 12

Addition and Subtraction

2 + 13 =	
8 - 6 =	
3 + 7 =	
4 - 3 =	
10 + 2 =	

15

2

10

1

12

Addition and Subtraction

8	+	4	=	
10	-	9	=	
14	+	4	=	
13	-	3	=	
4	+	3	=	

12
1
18
10
7

Addition and Subtraction

5	+	5	=	
15	-	3	=	
12	+	2	=	
19	-	8	=	
16	+	2	=	

10
12
14
11
18

Addition and Subtraction

3	+	9	=	
14	-	10	=	
4	+	14	=	
16	-	7	=	
2	+	3	=	

- 12
- 4
- 18
- 9
- 5

Bonus Activity

Learning numbers

1 1 1 1 1 1 1 1 1 1 1
2 2 2 2 2 2 2 2 2 2 2
3 3 3 3 3 3 3 3 3 3 3
4 4 4 4 4 4 4 4 4 4 4
5 5 5 5 5 5 5 5 5 5 5
6 6 6 6 6 6 6 6 6 6 6
7 7 7 7 7 7 7 7 7 7 7
8 8 8 8 8 8 8 8 8 8 8
9 9 9 9 9 9 9 9 9 9 9
10 10 10 10 10 10 10 10 10 10 10

HOW MANY?

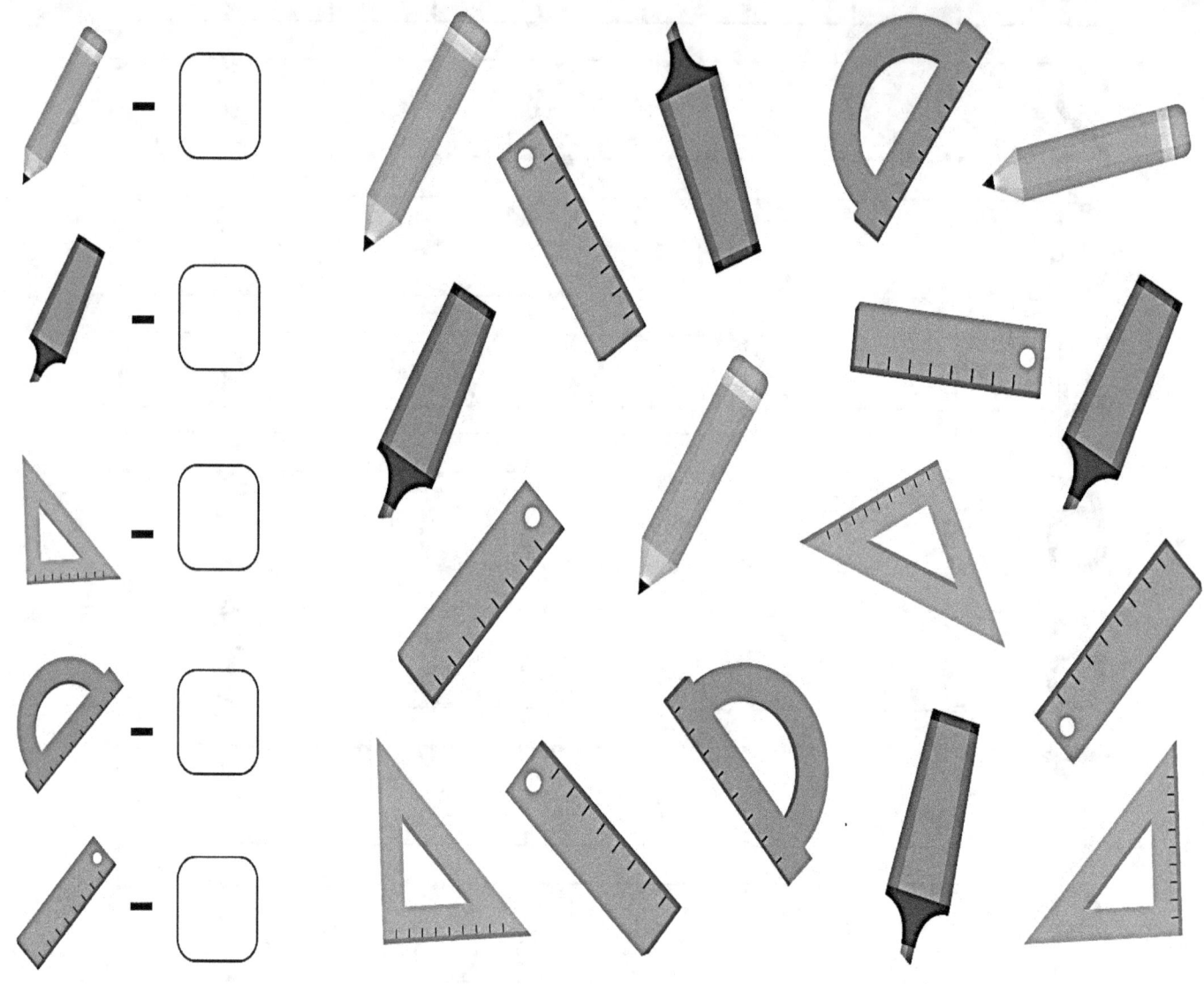

HOW MANY?

 1 2 3 4 5

HOW MANY?

 3 4 5 6 7

ANSWER
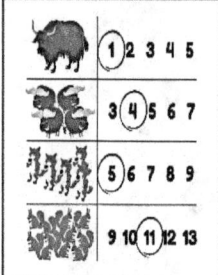

 5 6 7 8 9

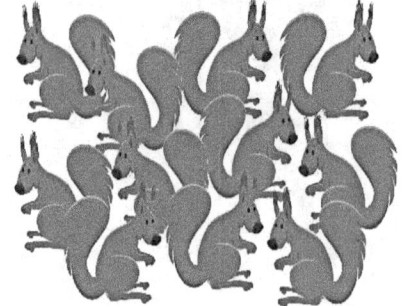 9 10 11 12 13

1 2 3 4 5

HOW MANY?

3 4 5 6 7

ANSWER

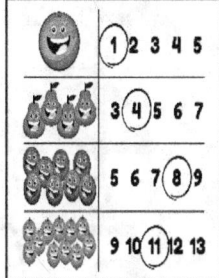

5 6 7 8 9

9 10 11 12 13

Learning numbers

1 1 1 1 1 1 1 1 1 1 1
2 2 2 2 2 2 2 2 2 2 2
3 3 3 3 3 3 3 3 3 3 3
4 4 4 4 4 4 4 4 4 4 4
5 5 5 5 5 5 5 5 5 5 5
6 6 6 6 6 6 6 6 6 6 6
7 7 7 7 7 7 7 7 7 7 7
8 8 8 8 8 8 8 8 8 8 8
9 9 9 9 9 9 9 9 9 9
10 10 10 10 10 10 10 10 10